Oxford Physics Series

General Editors
E.J.BURGE   D.J.E.INGRAM   J.A.D.MATTHEW

Oxford Physics Series

R.L.F. BOYD

MULLARD SPACE SCIENCE LABORATORY, UNIVERSITY COLLEGE, LONDON

Space Physics

the study of plasmas in space

Clarendon Press · Oxford · 1974

Oxford University Press, Ely House, London W.1

GLASGOW   NEW YORK   TORONTO   MELBOURNE   WELLINGTON
CAPE TOWN   IBADAN   NAIROBI   DAR ES SALAAM   LUSAKA   ADDIS ABABA
DELHI   BOMBAY   CALCUTTA   MADRAS   KARACHI   LAHORE   DACCA
KUALA LUMPUR   SINGAPORE   HONG KONG   TOKYO

CASEBOUND ISBN 0 19 851807 2

PRINTED IN GREAT BRITAIN BY
J. W. ARROWSMITH LTD., BRISTOL, ENGLAND

Editors' foreword

SPACE physics can cover a wide variety of topics and conjure up in the mind a whole range of applications spreading into engineering, geology, and many other fields. This text is concerned not with the wide range of specialized applications, however, but rather with the basic *physics* to be associated with space and its properties. In order to encompass a coherent presentation in one small volume a selection even from within the field of basic physics has to be made, and as Professor Boyd explains in his introduction, he has concentrated on the physics of matter filling most of space—i.e. *plasmas*— and left the consideration of the solid state to later books in the series.

It is probably somewhat surprising to most of us that the universe is composed almost wholly of plasmas, but as more detailed investigations take place this becomes increasingly evident—the solar wind and corona, the material expanding into space from supernova explosions, and the regions around neutron stars and black holes being a few examples. It is clear, therefore, that any understanding of the physics of space will need, as part of its foundation, an understanding of the nature and properties of plasmas and of the methods used to study them—and it is just such an understanding that this book seeks to give.

The different types of plasma that can occur are taken up in the different sections of the text, the first chapter dealing with 'ionospheres', including their discovery and properties, while the second chapter moves on to 'magneto-spheres', and starts with the discovery of the earth's radiation belts by the first American satellite that was launched. Throughout the text there is a careful balance between theory and experiment—and in a subject like this the experimental methods are in the forefront of modern technology, with satellite probes playing a major part.

The Oxford Physics series has been particularly fortunate in securing Professor Boyd as an author on this topic since he himself has pioneered the use of satellite probes in this country and can write with a personal authority on them which is evident from his fascinating accounts of their use. This book is one of the optional texts of the series and is intended to fit into a course at the end of the first year, or during the second year of studies. Thus, although the topics covered by this text are not reckoned as essential to every course in physics, the book has nevertheless been designed to fit in with the core volumes of the series on *Electromagnetism, D.c. and a.c. circuits, Radiation and quantum physics, Atoms and their structure, Atoms in contact, Interactions of particles*, and *Nuclear physics*. These core texts relate closely to each other and lead on to second- and third-year topics in quantum mechanics, statistical mechanics, solid state, and surface physics. This particular volume also links on directly with a group of texts in astronomy which are being prepared, and in this way the whole series is designed to reflect and match the more flexible nature of the new physics courses that are being designed at the moment to give variety of approach but, at the same time, an integrated and coherent picture.

D.J.E.I.

Preface

SINCE the launching of the first artificial Earth satellite in 1957 the word *space* has become an adjective commonly used to denote activities involving *space*craft. Space research, space physics, space chemistry, space biology, and so on refer to studies made possible by rocket propulsion although astronomy, astrophysics, lunar chemistry, and even some theoretical planetary biology all have a longer history. It is in the sense of depending on rocket techniques that *Space physics* is employed as the title of this book.

The popularity of that most ancient branch of applied physics, astronomy, has increased greatly of recent years no doubt partly because of the stimulus of space research and partly because it at least seems to embrace the glamour and mystery inherent in a fundamental science while offering no threat to mankind either through misuse of the environment or from human strife.

The Oxford Physics Series is to include a number of texts aimed to make available to the undergraduate early in his studies, an authoritative account of several topics of a generally geophysical, astrophysical, or cosmological character so as to aquaint him with some of the exciting developments and imperious questions being actively pursued in his subject. This book is concerned with researches in which space techniques have played a major part, other than those concerned with solid systems (e.g. the Moon and the planets). As the Introduction makes clear, elimination of solid systems leaves by far the greater part of the Universe, both in terms of mass and of space, and nearly all of this material exists in the fourth state of matter—an electrical plasma.

The material in this book is ordered working from the Earth's upper atmosphere outwards. In approaching the subjects in this way we move from a region already extensively studied before the advent of the research rocket to the magnetosphere, whose existence was barely recognized and the delineation of whose properties has been carried out only since the start of space research and mostly by its techniques. Leaving the environment of the Earth we turn to consider the Sun, upon whose behaviour that environment so largely depends. Not only is the subject of solar–terrestrial relations an important link between studies of the Sun and the Earth, but the behaviour of their plasmas in the presence of magnetic fields also corresponds in many ways.

From the Sun we move out again to our galaxy and far beyond to systems quite unknown before the start of astronomy from space vehicles. Here we find that the view of the Universe beyond the obscuring curtain of our atmosphere gives an insight into a variety of phenomena of great cosmological and astrophysical significance: interstellar shock waves from remote supernova explosions, rapid regular and secular variations of X-ray fluxes from neutron stars, with a density like that of nuclei, and the departing footprint of 'black holes'. Yet again in these events where matter is so different from its terrestrial form and where gravity, the weakest of all interparticle forces, holds sway we find the behaviour of plasma in magnetic fields is a controlling influence.

R.L.F.B.

Acknowledgements

THE author wishes to express his gratitude to Dr. D. J. E. Ingram a general editor of this series for his helpful comments at all stages in the preparation of the text, to the Clarendon Press for an efficient and helpful approach, to colleagues upon whose work he has drawn, and to authors and publishers of several of the figures reproduced here.

Fig. 2.13 is reproduced by permission of the Max–Planck Institut für Physik und Astrophysik and the Pergamon Press. Permission to reproduce Figs. 3.3 and 3.4 was given by Professor R. Wilson and The Royal Astronomical Society. Dr. Newkirk and the High Altitude Observatory, Boulder gave permission to use Fig. 3.6. NASA permitted the use of Figs. 3.12, 3.14, and 3.16. Dr. D. E. Osterbroek, director of the Lick Observatory granted permission to use Fig. 4.3 and Figs. 3.15, 4.1, 4.2, and 4.4 were supplied by Dr. R. Giacconi of the Smithsonian Astrophysical Observatory. Permission to reproduce Fig. 3.15 was given by the *Astrophysical Journal* (University of Chicago Press).

Contents

Introduction

Scope of the book

W HEN we come to study space we find that the Universe is mainly composed of that strange form of matter called *plasma* rather than normal fluid or solid materials. Any book on space physics therefore must be largely concerned with plasmas and must give consideration to the physics of this state of matter in some detail. It is for this reason that the subtitle of this book is *The study of plasmas in space*, and since a book of this length can only deal with certain selected subjects, the physics of the solid planets has been omitted. The general plan of the book has been to select four very different regions of cosmic plasma, to discuss some of their basic properties and to show how space techniques have been, and are being, applied to them. Since the idea of a plasma may well be novel, a brief general introduction to the subject follows before we discuss some different types of plasma more thoroughly in the succeeding chapters.

Plasma

In 1929 Irving Langmuir adopted the term plasma to describe the fluorescent gas in an electrical discharge. The word comes from the Greek for 'something moulded' and aptly describes the way in which the glowing gas fills the shape of the tube and responds to applied fields almost like a living thing. So different is the behaviour of a plasma from ordinary solid, liquid, or gas that it is sometimes referred to as the fourth state of matter.

The special behaviour arises from the fact that an electrical plasma consists of huge but very nearly equal numbers of charged particles of opposite sign. Each particle is surrounded by its Coulomb electric field, which, because this is a very long-range force compared with van der Waals interparticle forces, implies that each charged particle exerts an influence on, and is influenced by, a large number of other particles. It is this property which gives a plasma its coherent behaviour, that makes it capable of oscillating somewhat like a jelly, and forges such a strong link between it and any interpenetrating magnetic field.

Plasmas in general are very good conductors of electricity and, because of the high mobility of electrons, are also good thermal conductors. As a result of their high conductivity they have the peculiar property of not being able to support appreciable d.c. electric fields (except normal to any magnetic field present). This comes about because a d.c. field causes a movement of charges towards or away from the electrodes producing the field, according to their sign. The *space charge*, resulting as positive charges move towards a negative electrode and negative charges move away from it (and vice versa), screens the plasma from the field and localizes it in a *sheath* around the electrodes.

Debye, in studying the theory of strong electrolytes which are themselves plasmas, showed that this sheath screening effect is not limited to electrodes but occurs around each individual charged particle so that its field is modified from the simple Coulomb form and limited to a region of the order of the so-called Debye length given by $69(T_e/n_e)^{\frac{1}{2}}$ m. where T_e is the electron temperature and n_e the number of electrons per cubic metre.

When there is a magnetic field present or when the electric field is transient or alternating, the situation is changed. We shall see in Chapter 2 that a magnetic field inhibits a flow of charged particles across it, so that significant electric fields can then arise in that direction. Moreover, a changing magnetic field induces large currents which tend to neutralize the change in the magnetic field, with the result that mutual motion between plasma and magnetic field is restricted.

Plasmas in space

The ionosphere, the magnetosphere, the interplanetary medium, the interstellar medium, and the intergalactic medium (if such there is) are all plasmas. The atmosphere of the Sun and stars, the material from supernova stellar explosions, the shock-wave excited media from cosmic explosions, the region around neutron stars and black holes, the mysterious jet from the galaxy M87 (see Fig. 4.3), and other similar, and indeed many dissimilar, cosmic phenomena are plasma. It is hardly surprising, therefore, that this book on space physics is so largely concerned with space plasma.

Since selection of material is essential besides restricting the book to space plasma, the author has drawn much from the research of the Mullard Space Science Laboratory of University College London, which is situated in a beauty spot in the Surrey hills. After all, if one has to select there is something to be said for selecting that of which one has direct experience. The underlying link that has always influenced the choice of the Laboratory's programme is that which relates and dictates the subjects of this book—what parts of space present us with interesting and exciting plasma phenomena amenable to the diagnostic techniques made available by spacecraft?

In ordering this work we move from ionospheric plasma near the Earth, which can be probed directly, to magnetospheric plasma, where the vital role of magnetic fields is introduced, and out to the Sun, which presents us with a huge intensely hot coronal plasma which is both more rarified than can be reproduced in the laboratory and sufficiently extensive for interaction between radiation and particles to be significant. Finally, we turn to a new and most exciting branch of astronomy, where plasmas at tens to hundreds of millions of degrees radiate many billions of megawatts and are heated by such exotic processes as conversion of gravitational potential energy, huge cosmic dynamos, or shock waves in space.

The origin of space plasmas

It is not our business here to discuss the detailed physics and chemistry of plasmas in space, but rather to describe how they may be studied and what are the highlights and gaps in the picture as we have it. Before such a study is undertaken we must get some understanding of the origin and nature of the phenomena.

Plasmas in space, as in the laboratory, result from one or more of three main production mechanisms. The first of these is thermal ionization as in the carbon arc, thermo-nuclear fusion experiments, and the outer atmosphere of the Sun and stars (we shall discuss this in Chapters 3 and 4). In the first laboratory case the energy input is ohmic in character, in the second it can be compression of the plasma, in the Sun's case it is dissipation of shock-wave energy, while in some X-ray stars it is gravity. A special laboratory case is that of the low-pressure arc plasma, where the high value of E/p (electric field/pressure) results in mean electron energies one or two orders of magnitude higher than those of the ions or neutral particles. In the laboratory it is usually the loss of ions to the walls which is responsible for the great disparity between the temperature of the charged species, and it is questionable whether comparable ratios of T_e/T_+ ($\sim 10^2$) occur in accessible space situations. It is very important to remember this when comparing theory and experiment of probe electrodes used in the laboratory with their behaviour in space. Moreover, in the low-pressure arc plasma, because of the spatial distribution of E/p imposed by the boundary conditions of walls, cathode, and anode, the energy distributions of the charged species are often far from Maxwellian in form. Departures of T_e/T_+ from unity do occur in space even in the absence of significant electric fields (for example, in the topside ionosphere during daylight $T_e/T_+ \sim 2$), and there may be discharge types of phenomena or regions of high E/p (connected, for example, with aurorae) where considerably higher values will be found. On the whole, however, it is often safe to assume that true temperatures exist for the ions and electrons in space plasmas, though they are not by any means always in thermal equilibrium with each other.

The second main plasma-production mechanism is that of energetic streams of particles, as in some electron-bombardment ion sources for mass spectrographs or isotope separators. There are many space examples of this process. The normal auroral and the polar-cap absorption ionization arise respectively from electron and from proton streams. The ionization in the lowest part of the ionosphere is produced by the primary cosmic rays and their energetic secondaries. A case in which both the initial components are uncharged is the phenomenon of formation of luminous and ionized meteor trails in which there is ionization of metal atoms, evaporated from meteors, on collision with atmospheric atoms and molecules. Instrumentation to study the ionosphere must take account of the possible effects of streams of particles. Of course, not only the ambient plasma but these particles may themselves be the object of the measurement (see Chapter 2).

The other main source of space plasma is photo-ionization, indeed in the ionosphere this is predominant. If the photons have an energy greater than twice the ionization energy of the gas with which they are reacting, the ejected electrons may themselves contribute to further ionization, and in any case any excess energy may result in heating. The production of energetic photo-electrons is especially significant when the electron mean free path is long, for then the effects to which they give rise may occur far away from the region where the photon is absorbed, perhaps even, in the case of the ionosphere, in the other hemisphere of the Earth. Because the Coulomb force acts over long ranges, the photo-electrons can exchange energy readily with ambient electrons and so raise their temperature above that of the ions and neutral particles. These processes are especially important in the topside ionosphere.

Methods of studying plasmas

Ionized regions where the free positive and negative charges cancel to a high degree are studied, whether in the laboratory or in space, by four main diagnostic techniques. These are:

(1) the examination of the electromagnetic radiation to which they give rise;
(2) the examination of particles leaving them;
(3) the study of currents to electrodes (probes) immersed in them;
(4) the study of their effects on electromagnetic radiation passing through them.

Of these methods (1), the examination of radiation from ionized regions, forms the subject *per force* of space astronomy and so of Chapters 3 and 4, although in Chapter 3 (2), the flux of particles from the solar plasma, is involved, since we can study particles leaving the solar plasma at the Earth and in interplanetary space. The study of radio propagation—method (4)—has been the historical approach to the ionosphere and to some extent to the magneto-sphere. The advent of spacecraft has enabled extensive use of this method, especially in studying the otherwise largely inaccessible topside ionosphere. Being able to put one or other, or both, of the transmitting or receiving ter-minals in a spacecraft greatly increases the flexibility of the method.

In this book we shall content ourselves with emphasizing the continuing importance of radio propagation in ionospheric and in magnetospheric studies, and will limit our discussion of ionospheric experiments to the direct probing techniques (2) and (3).

In the case of the ionosphere and magnetosphere the distinction between studying particles leaving the plasma and studying currents to electrodes immersed in the plasma is somewhat blurred. There is however an important practical consideration which can mark the difference. Particles that have energy large compared with that represented by the difference in potential between the spacecraft and its surrounding plasma can be thought of as in category (2). For these the energy (and velocity) of arrival at some sampling orifice on the craft is scarcely dependent on the potential of the spacecraft. In category (3) are the thermal distributions of charged particles around the craft. These have a controlling influence on its potential, and their fluxes are therefore strongly dependent on the potential of the sampling electrode system.

1. Ionospheres

Discovery of the ionosphere

THE first cosmic ionized medium or plasma to be recognized was, naturally enough, the terrestrial ionosphere, and although we will mention the ionospheres of other planets which are now becoming accessible to direct probing methods, it is the envelope of ionized gas around the Earth that has been most studied and is the main concern of this chapter. The ionosphere's existence was postulated in 1882 by the Scots physicist Balfour Stewart, in an article in the famous ninth volume of the Encyclopaedia Brittanica, and it is interesting to note how broad his predictions were, and how correct.

The basic evidence from which Stewart reasoned was the clear connection between the Sun and geomagnetism, shown by the fact that the regular diurnal variations of the compass are greater by 50 per cent at sunspot maximum over the excursions at sunspot minimum. To account for this he postulated (1) that the magnetic variations are due to electric currents in the high atmosphere; (2) that the upper air is rendered conducting by the action of the solar radiation upon it; (3) that the currents are induced by the dynamo action of the region's motion across the geomagnetic field as a result of global winds; (4) that the atmosphere is more highly conducting during the day, in the summer hemisphere, and at sunspot maximum.

It was, however, twenty years before Marconi's celebrated transatlantic radio transmission, on 12 December 1901, provided fresh and irrefutable evidence for the existence of the ionosphere. Heaviside and Kennelly drew attention independently to this corollary of Marconi's experiment and Kennelly wrote: 'As soon as long-distance wireless waves come under the survey of accurate measurements we may hope to find ... data for computing the electrical conditions of the upper atmosphere'.

These 'accurate measurements' were awaited for another twenty-two years, until the pioneering work of Appleton in the U.K. and Breit and Tuve in the U.S.A. Appleton concentrated on interferometric studies of ionospheric reflections using continuous-wave transmission, and so was concerned with the path of the ray measured in wavelengths λ, while the American workers used pulses of radio transmission and measured the travel time of a pulse or wave packet moving at a speed c, a technique which nowadays we should associate with radar. In each case the frequency $f(=c/\lambda)$ of the probing wave was an important parameter, since in the ray theory reflection occurs when the sine of the angle of incidence on the reflecting layer is equal to the refractive index of the layer (just as total internal reflection in optics occurs in passing from a medium of refractive index μ_1 to one of index μ_2 when sin $i =$

μ_1/μ_2). The refractive index is related to the local electron concentration n_e by†

$$\mu = \left(1 - \frac{e^2 n_e}{\varepsilon_0 m \omega^2}\right)^{\frac{1}{2}}, \tag{1.1}$$

where ε_0 is the permittivity of free space, m is the mass of the electron, and ω the angular frequency of the wave.

It was soon discovered that there were more than one reflecting layer of ionization in the upper atmosphere and Appleton suggested a nomenclature using the letters of the alphabet, the original Kennelly–Heaviside layer which occurs at a little over 100 km altitude being designated the *E layer* and the higher layer, sometimes called the Appleton layer, being referred to by the letter F. Wisely, Appleton started at E, the symbol he had used to designate the electric vector of the wave there, in case lower layers should be discovered : later a *D layer* and even a *C layer* were added, although these latter are perhaps better referred to as *regions* of ionization, because the true layer structure, which we shall discuss later, is not pronounced. It was also discovered that the *F layer* split into two reflecting regions at different altitudes during the day so the lower was labelled the F_1 *layer* and the upper the F_2 *layer*. The region beyond the ionization maximum of the F layer is now often referred to as the *topside ionosphere*. Beyond the ionosphere is a region where the Earth's magnetic field rather than its gravitational field exerts a controlling influence. This is the *magnetosphere* and is the subject of Chapter 2. It may be helpful here to glance at Figs. 1.3, 2.5, and 3.6.

Further away still, beyond a region of transition, is the interplanetary plasma which is contiguous with and in some ways indistinguishable from the *corona*, or outer atmosphere of the Sun, which we shall discuss in Chapter 3.

Limitations of ground-based observations

As long as indirect probing by radio signals from the ground was the only way of exploring the ionosphere, serious gaps in our knowledge were inevitable. The radio-probing methods exaggerated the layer-like structure and led to the adoption of theoretical models with very clear troughs of ionization between the peaks. A glance at an ordinary *ionogram* (see Fig. 1.1), as the records of pulse travel time versus frequency from a vertical incidence *ionosond* are called, shows how this conclusion was natural. The heights shown directly by the pulse travel times on these ionograms are known as *virtual heights*. They are the altitude given by assuming that the wave packet travels with the speed of light. But, in fact, the phase velocity of the waves is increased to c/μ and the group velocity decreased to μc, so that the true height will always

---

† This is a simplified form of the equation which neglects the effect of the Earth's magnetic field. The field splits the refractive index into two different values for oppositely circularly polarized waves.

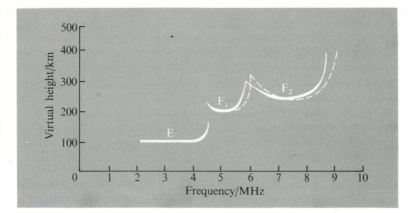

Fig. 1.1. Ionogram showing E- and F-layer reflections. Extraordinary ray dashed.

be less than the virtual height, since the packet must traverse some plasma before reaching the *critical density* which results in reflection. Clearly, the difference between true and virtual heights depends on the distance travelled by the wave packet in the ionosphere before reaching the critical density and also on the value of the refractive index over this path. It is this rapid fall in group velocity as a *critical frequency* (a frequency for which the refractive index falls to zero) is approached that leads to the curious cusp-like structure of the ionogram. The other prominent feature, the splitting of the diagram into twin curves, is due to the birefringence of the plasma (see footnote to p. 2) resulting from the presence of the Earth's magnetic field. Sometimes echoes appear also, which suggests much higher layers, but these are due to pulses making a second journey after reflection back from the ground.

Just at the time rocket-probing of the ionosphere was being introduced, techniques were worked out for solving the integral equation for the pulse delay of a given frequency in terms of the profile of pulse delays at lower frequencies given by the ionogram. In this way, given an ionogram of exceptional quality, reasonable values of true heights for the layer maxima could be obtained, together with some information on the existence and depth of any trough between layers. Nevertheless, it remains true that, below the F layer maximum, accurate height values and detailed density profiles require the use of rocket-probing techniques. The same general conclusion holds for the topside ionosphere. Here, very high-power *incoherent-scatter radar* has made it possible, at the few sites where such equipment exists, to obtain electron concentration profiles above the F layer maximum and also to extract data on electron and ion temperatures. The term 'incoherent scatter' implies that the reflections arise from statistical irregularities in the electron concentration.

These developments have been contemporaneous with those of direct probing by instruments on spacecraft and the making of topside ionograms from satellites looking downwards, but the advantage of spatial resolution and accuracy remains with the direct probing techniques.

Apart from the very limited applicability of incoherent-scatter radar, it can be said that the measurement of electron and ion temperatures, which are so important if we are to understand the way the Sun's energy flows into and out of the ionosphere, is dependent on spacecraft. The measurement of ionic composition is vital when we try to interpret the photo-electric and photochemical reactions going on and is almost wholly dependent on such techniques. Where precise data are required with good spatial resolution they are seldom obtainable without direct probing. Rocket research is expensive however, and these conclusions make a case for a close correlation between ground and rocket methods. The greater understanding coming from the direct study illuminates the data obtained from radio propagation, whether ground-to-ground or between spacecraft and ground, and this data itself helps to fill the temporal and spatial gaps in direct probing.

The formation of an ionized layer

The basic theory of the formation of ionospheric layers was worked out originally by Chapman in 1931.

The number of particles n per unit volume of the neutral atmosphere falls with altitude z according to the equation of hydrostatic equilibrium, as can easily be shown by considering the equality of the total forces acting upwards and downwards on a stationary elementary slab of atmosphere. This equation is

$$n_z = n_0 \exp -(z/H), \tag{1.2}$$

where $H = kT/\bar{m}g$ and is known as the *scale height*. k is Boltzmann's constant, T is the temperature, $\bar{m}$ is the mean molecular mass, and g is the strength of the gravitational field. T, $\bar{m}$, and g all vary with height, but often sufficiently slowly for the hydrostatic equation to be integrated over a substantial range with the approximation $H = $ constant. We shall see later that it is often necessary to consider the molecular masses of the individual constituents separately.

Sometimes allowance is made for the inverse square law variation in g by changing the height variable to z' the *geopotential metre*. Altitude is then measured linearly in terms of units of gravitational potential difference equal to the gravitational field at the Earth's surface. It is easy to show that

$$z = \frac{R_0}{R_0 + z'} \simeq \left(1 - \frac{z'}{R_0}\right), \tag{1.3}$$

where R_0 is the radius of the Earth (from which z is measured) and at which $g = 9.81 \text{ m s}^{-2}$.

In the idealized case of an isothermal single-constituent atmosphere in hydrostatic equilibrium above a flat Earth (g constant), Chapman's theory shows that a layer of ionization will be produced, as Heaviside had supposed. This is only to be expected, since the rate of production of ion–electron pairs will clearly increase as the Sun's radiation penetrates deeper into an atmosphere of ever increasing density, until the flux itself begins to be significantly absorbed and finally is attenuated away to nothing. Chapman's theory gives the rate of ionization q_h as

$$q_h = Q \exp\{1 - h - \sec\chi \exp(-h)\}, \tag{1.4}$$

where the height variable h is defined in terms of the height measured in units of H from the altitude Z of maximum ionization production for a zenith Sun. Thus

$$h = (z - Z)/H. \tag{1.5}$$

χ is the solar *zenith angle*, that is, the angle between the Sun's direction and the vertical, and Q is the maximum rate of ionization for a zenith Sun and is given by

$$\eta I/H \exp(-1). \tag{1.6}$$

η is the ionizing efficiency of the radiation in electrons produced per photon absorbed (if λ is the wavelength in nanometres, $\eta \sim 36/\lambda$, except near the ionization threshold).

The altitude Z at which the ionization production rate is a maximum is given by

$$Z = H \ln(\sigma n_0 H), \tag{1.7}$$

where n_0 is the neutral particle concentration at $z = 0$ and σ is the cross-section for photon absorption.

Ionospheric experiments do not observe the rate of production of ions directly but measure n_e or n_+, the resulting electron or ion concentration. If the ion loss rate follows a recombination law ($\mathrm{d}n_e/\mathrm{d}t = \mathrm{d}n_+/\mathrm{d}t = -\alpha n_e n_+$), with coefficient α, at equilibrium (which may not necessarily persist for a finite time),

$$n_e = (q/\alpha)^{1/2} \quad \text{or} \quad N_e = (Q/\alpha)^{1/2}, \tag{1.8}$$

where N_e is the maximum equilibrium electron concentration for a zenith Sun. Eqn (1.4) leads to

$$n_e = N_e \exp\tfrac{1}{2}\{1 - h - \sec\chi \exp(-h)\}. \tag{1.9}$$

For small values of h this is parabolic in form

$$n_e \sim N_e\{1 - (h/2)^2\}, \tag{1.10}$$

while for large values of h it represents an exponential fall of density with altitude, at half the rate of fall of the neutral particle density. That is, H_e the scale height for the electrons is given by $H_e = 2H$ (see Fig. 1.2 and p. 26).

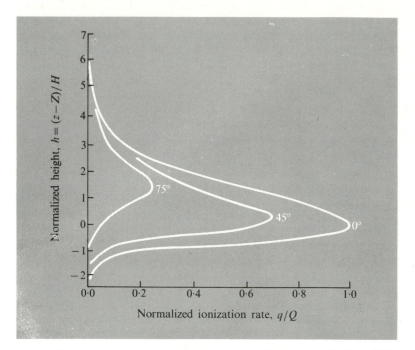

FIG. 1.2. Rate of production of ionization for solar zenith angles $\chi = 0°, 45°$, and $75°$. Altitude h is in units of scale height measured from the height of the maximum for a zenith sun ($\chi = 0$). Ionization rate is normalized to unity for the same maximum. (After Chapman 1931.)

In any real planetary situation, and especially in the case of the Earth, the atmosphere does not consist of a single gas, and the solar spectrum contains many spectral lines and some continua capable of ionizing the various components. There are, therefore, many values for σ and η and hence for Z, and it is therefore necessary to integrate over the solar spectrum to find the total effect. Moreover, ion-loss processes are complex and may vary substantially from a simple recombination law; chemical reactions change the nature of the ions and, at the top of the ionosphere, recombination becomes less important than the motion of ions under the influence of hydrostatic, magnetic, and diffusive forces, so that the distribution of ionization is controlled more by these than by a production-loss equilibrium. The observed distribution of ionization in the photo-electrically produced ionosphere therefore may depart substantially from that given by a simple Chapman theory, and as we would expect these departures have been the stimulus behind a number of experimental researches.

In addition to the broad structural features, there are irregularities many of which are produced by magneto-hydrodynamic forces and/or the effects of corpuscular ionization. Prominent amongst these are the equatorial, mid-latitude, and auroral *sporadic-E* phenomena, where layers of ionization are found in the E region much denser and often much more shallow than Chapman's theory would predict, and the *field-aligned irregularities* of the topside ionosphere, in which the Earth's magnetic field clearly plays an important role.

The solar ionizing radiation

The atmosphere of the Sun is the subject of Chapter 3, but since the ionosphere is an important aspect of solar–terrestrial relations something about the Sun must be said here.

Until the advent of space techniques little was known about the distribution of energy in the solar spectrum in the ultraviolet and X-ray region, though it had been inferred that the energy distribution departed significantly from that of a black body at about 6000 K—the apparent temperature of the Sun in visible light—and might be expected to contain strong *emission* lines. From 1946 onwards, the use of space techniques has shown that, by the time wavelengths sufficiently short to ionize atmospheric constituents have been reached, the famous Fraunhofer spectrum of dark (*absorption*) lines has given place to a predominantly emission-line spectrum. Much of the energy is concentrated in the *resonance* lines of hydrogen Ly-α (121·6 nm) with a flux of about 5×10^{-3} J m^{-2} and helium Ly-α (30·4 nm) with a flux some twenty times less. There are, however, numerous other lines, principally from multiply-ionized elements with strengths about an order of magnitude less again.

The ultraviolet, and extreme ultraviolet (shorter than, say, 60 nm) spectrum comes mostly from the solar *chromosphere*, an irregular and dynamically disturbed region of the Sun between the visible *photosphere* and the immensely hot *corona* or outer atmosphere. In these wavelengths the Sun is a variable star. While the fluxes from a given area of the Sun show a considerable variability, ultraviolet and extreme ultraviolet energy arriving from the whole solar disc at the top of the atmosphere varies by less than a factor 2 over the solar sunspot cycle or with occurrence of solar flares. On the other hand, the X-ray flux, which is conventionally thought of as shorter than 10 nm, is much more variable. Beyond 1 nm this variation can exceed two orders of magnitude, which is noteworthy when it is realized that this implies that during a major flare, the X-ray emission between say 0·2 nm and 0·8 nm from the relatively small disturbed region can exceed that of a hundred suns. No wonder solar flares have a most disturbing effect on the ionosphere and so on radio reception.

The ionospheric regions

The terrestrial ionosphere is usually thought of as 'the part of the Earth's upper atmosphere where ions and electrons are present in quantities sufficient to affect the propagation of radio waves'†, a region which, under disturbed solar conditions, may start as low as about 40 km, that is, near the ceiling height for balloon studies. We must not think, however, that the ionization density necessarily decreases below this, but rather that the rising neutral particle concentration tends to the elimination of electrons by their attachment to molecules to form relatively immobile negative ions, whose effect on radio waves is marginal. Fig. 1.3 shows the approximate variation of electron concentration with altitude at mid-latitudes.

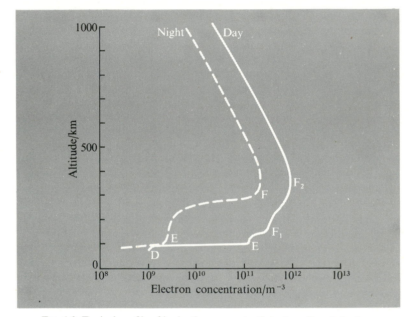

FIG. 1.3. Typical profile of ionization concentration at medium latitudes.

At the COSPAR meeting in 1967 there was a report of the observation by rocket of a broad peak in daytime electron concentration over White Sands, New Mexico, at 53 km under somewhat disturbed solar conditions, but the ion density continued to rise with falling altitude below this. There was also a report of a maximum in the positive-ion concentration at between 15 km and 25 km.

† Note that this definition makes the magnetosphere part of the ionosphere. Usage varies on this point.

This *C region*, as the ionosphere between 40 km and 60 km is now sometimes called, is due to cosmic rays ionizing the atmosphere constituents, and so may be expected to show a related latitude-dependence and a correlation with solar activity corresponding to the cosmic-ray correlation. In the auroral zones, ionization in this range may be caused by particles of magneto-spheric or solar origin. The nature of both positive and negative ions here is still uncertain, and the study of the region is only now growing in momentum. The greater glamour of greater heights may be partly responsible for the relatively little attention paid to the lowest parts of the ionosphere in the early years of space research, but undoubtedly the technical difficulty carries a large share of the responsibility. This difficulty arises pre-eminently from the short mean free paths of the ions and neutral particles and from the low ion and even lower electron concentration, which makes measurements very difficult. There is a need for the development of methods to study electron concentration and ion-mass spectra in this region.

The *D region* is defined as that between 60 km and 90 km. It has been exten-sively studied by radio-propagation techniques, because of its importance as a reflector of very low-frequency radio waves and attenuator of higher frequencies. Ionization in the region is due to the absorption of the solar hydrogen Ly-α (121·6 nm) radiation by nitric oxide and to absorption in all the atmospheric constituents of X-rays shorter than 1 nm. In periods of solar activity the latter process predominates. Until very recently ignorance of the ions of the D region was complete, but there is some evidence from mass spectrometers carried on rockets for NO_2^- as the major negative ion with NO^+ and O_2^+ as the major positive ions. Electrons are present to the extent of a few per cent of the negative ions in the daytime at the bottom of the region, owing to photo-detachment. Near the top of the region during the daytime, the electron concentration begins to approach that of the negative ions. There is need for more ion-mass spectrometry by differing techniques to enable cross-checking, but in some ways the problems are even more severe than in the C region, because the densities are transitional between those permitting diffusion-coefficient studies with parachute-braked subsonic systems and those permitting collisionless analysis using hypersonic rockets. In both C and D regions electron and ion temperature are expected to be close to that of the neutral gas.

The *E region*, stretching from 90 km to 160 km, has been studied extensively by sounding rockets. The mean free path is long enough to allow molecular flow conditions to be assumed, while ion densities around $10^{11}\,\mathrm{m}^{-3}$ give good conditions for measurements by probing electrodes. The major ion has been shown by mass spectrometry to be NO^+, formed mostly by ion–atom interchange betweed O^+ and N_2 (see following discussion of F region). Oxygen molecular and atomic ions also occur in substantial amounts, the latter beginning to predominate at the top of the region. Negative ions may be present in significant amounts at night, but measurements are needed here, though they are technically very difficult because the fluxes to a measuring system are so readily swamped by fluxes of the more mobile electrons.

The E region is marked by important irregularities which often reach concentrations greater than the noon maximum for the normal layer and which may occur over very short altitude distances (~ 1 km). These *sporadic-E* layers are, at least sometimes, dynamical in origin, being associated with the interaction of the wind structure with the geomagnetic field. Their thinness was revealed by rocket studies (see Fig. 1.4).

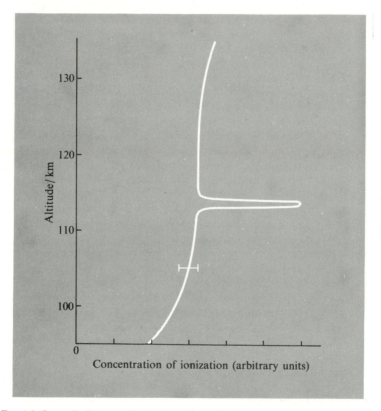

FIG. 1.4. 'Sporadic-E' layer of ionization observed by Skylark rocket fired at Woomera. Summer, daytime, 1961 (Wrenn, Willmore, and Boyd 1962).

The lower part of the region receives a considerable deposit of evaporated meteoric material, and layers of metallic ions have been observed by ion-mass spectrographs. The currents which give rise to the regular magnetic variations flow predominantly here and possibly may be associated with observable ionic or electronic concentration or temperature structure.

Because kinetic-energy exchange between particles is only efficient if they are of comparable mass, the electron temperature in the upper part of the region begins to rise well above the ion temperature, which remains fairly closely coupled to the neutral gas. This has been well studied by satellite instruments; however, in the lower E region, temperature data is still conflicting, and its correlation with sporadic-E and with magnetic phenomena is an important continuing study. The same applies to temperatures in the auroral regions, where particle fluxes are important. The major fluxes responsible for the E ionization seem to be the soft X-rays and the ultraviolet wavelengths capable of ionizing atomic and molecular oxygen but not nitrogen. The more energetic ultraviolet radiation is absorbed higher up, because of its greater absorption cross-section, and is responsible for the F layer.

The *F region* extends upwards from 160 km. The division from the E region is arbitrary, the main distinction being the greater importance of more energetic photons in its production and the falling rate of formation of NO^+ because of the rapid decrease in concentration of N_2 with height. Above 100 km the hydrostatic equation must be applied to each constituent separately, because of inadequate turbulent mixing. Photo-dissociation of nitrogen molecules proceeds slowly, and the large mass of the nitrogen molecule compared to that of the oxygen atom formed by photo-dissociation rapidly leads as we ascend to a neutral atmosphere consisting largely of atomic oxygen. Ion loss in the F layer proceeds as in the E layer mostly by dissociative recombination of NO^+,

$$NO^+ + e \rightarrow N + O.$$

Here the conservation of both energy and momentum is satisfied by the fact that two neutral particles result. In the F layer, however, the actual loss rate is determined by the rate of conversion of oxygen ions to nitric oxide ions by ion–atom rearrangement

$$O^+ + N_2 \rightarrow NO^+ + N,$$

a process which, unlike true recombination, varies in rate linearly with ionization concentration. Because of the smaller recombination rate the F layer persists at night, though the contribution of ionization flowing in from the magnetosphere cannot be discounted.

During the day, the F layer is bifurcated, the peak of the F_1 being a little below 200 km and the peak of the F_2 a little above 300 km, the actual altitude showing a dependence on season, latitude, and time of day but also giving evidence of layer movements due to effects other than those due to the solar zenith angle (tidal and magnetic forces). The maximum ion concentrations in the E and F region vary by a factor of about 2 and 4 respectively over the sunspot cycle, the mean value of the F_1 peak concentration being above 3×10^{11} m^{-3} and of the F_2 peak concentration being nearly 10^{12} m^{-3}.

The bifurcation of the F region is an outcome of the falling rate of loss of O^+ with altitude, indeed, since the scale height for the production of ionization q is approximately the scale height of oxygen atoms, while the scale height for its loss is more nearly the scale height of nitrogen molecules, the concentration of ions might be expected to continue to increase with height above the altitude of peak of production. The occurrence of a well-marked maximum for the F_2 layer, however, arises from the increasing significance, as the neutral atmosphere dwindles of downward diffusion of ions to a region where they can be converted to NO^+ and so rapidly disappear by dissociative recombination.

The region above the F_2 maximum, the *topside ionosphere*, is thus a region in which the profile is determined by the forces on the plasma fluid more than by the local production and loss equilibrium.

The ionospheres of Mars and Venus

The giant planets have atmospheres and, presumably, ionospheres, which will someday be explored by direct probing, though the ionization densities will correspond to the much smaller intensity of the ionizing flux at their distance. At the time of writing, Pioneer 10 has passed Jupiter on its way to the outer planets, and Pioneer 11 has been launched. The terrestrial planets Mars and Venus have well-developed ionospheres which have already been the subject of space-vehicle studies. Some star occultation experiments have suggested a weak ionosphere above the lunar surface.

Data on the ionospheres of Mars and Venus were obtained by the flights of Mariner IV (Mars) and Mariner V (Venus). These craft passed behind the planets and made possible a study of the refractive index of the ionospheres and atmosphere from the phase change and bending of the rays. The Mariner IV data showed a maximum ionization density of 10^{11} m^{-3} at above 120 km altitude, with a hint of a second layer at 95 km on the sunlit side of the planet—solar zenith angle 67°. Dark-side measurements—solar zenith angle 104°—gave no detectable ionization, which should probably be construed as implying that the electron density was at least an order of magnitude less. The electron scale height above the maximum was about 30 km. If CO_2^+ is taken to be the principal ion this would give a mean plasma temperature of 300 K.

The Mariner V measurements on Venus yielded the data of Fig. 1.5. The cytherian ionosphere appears to show a sudden decrease or plasmapause at above 500 km on the sunlit side, while on the night side the ionization streams out for thousands of kilometres. No doubt the absence of any significant magnetic field for the planet enables the solar wind to penetrate thus closely to the ionospheric layer instead of being stopped at a magnetospheric shock wave, as in the terrestrial case (see Chapter 2).

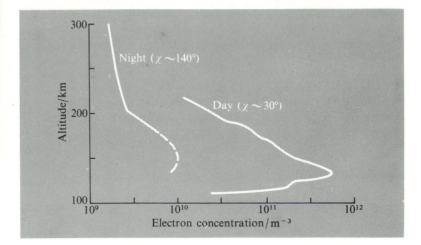

Fig. 1.5. Profiles of electron concentration in the mid-latitude ionosphere of Venus obtained by Mariner 5 (19th October 1967) using the effect of the refractive index of the medium on the Doppler shift in a transmitted wave due to the motion of the transmitter. (After Fjeldbo.)

The potential of a spacecraft

Because the potential of a spacecraft depends on the properties of the plasma through which it is moving and is frequently an important parameter in the behaviour of plasma probes, we must discuss it before considering the measurement techniques.

Think of a sphere r m diameter moving at 8 km s^{-1} through a plasma of density $m_+ = m_e = 10^{11}$ electrons and ions per cubic metre. If the ions were mostly O$^+$ at a temperature of 1600 K, which would be characteristic of the ionosphere at 400 km altitude, their mean thermal velocity $\bar{v}_+$ (say) would be about 1·3 km s^{-1}. The electrons which might have a temperature of 2500 K would have, as a result of their small mass, a much higher mean thermal velocity $\bar{v}_e = (8kT/\pi m)^{1/2}$. The velocity here differs slightly from that given by $\frac{1}{2}mv_{rms}^2 = \frac{3}{2}kT$, because the r.m.s. and mean velocities for a Maxwellian distribution are not quite the same. The mean velocity of the electrons would be about 300 km s$^{-1} = \bar{v}_e$ (say).

Now other things being equal the current of ions swept up by the sphere would be roughly

$$i_+ = qn_+\bar{v}_s\pi r^2, \tag{1.11}$$

since the velocity of the spacecraft v_s is so much greater than that of the ions. On the other hand, as the reverse is the case for the electrons their current would be the random flux to a sphere

$$i_0 = \tfrac{1}{4}qn_e\bar{v}_e 4\pi r^2 = qn_+\bar{v}_e\pi r^2. \tag{1.12}$$

This is greater than i_+ in the ratio $\bar{v}_e/\bar{v}_s$. It is impossible for an isolated sphere to continue to take a net current from the plasma so—in a fraction of a second—it will be charged negatively to a point which will limit the flow of electrons to equal that of the ions. Integration of a Maxwell velocity distribution or consideration of Boltzmann's relation shows that the current i_e of particles from such a distribution at a temperature T_e, which are able to climb a potential barrier V, is related to the current i_0 in the absence of such a potential hill by

$$i_e = -i_0 \exp -(qV/kT). \tag{1.13}$$

Here qV is the energy required to overcome a potential V, and we must remember to take q negative for electrons. In fact, it is better to retain q as the magnitude of the electronic charge and to adjust the sign in the relevant equations. Equating (1.11) and (1.13), with (1.12) substituted for i_0, to find the equilibrium potential of the satellite, we obtain:

$$\text{net current to satellite} = 0 = i_+ + i_e = qn_+\bar{v}_s\pi r^2 - qn_+\bar{v}_e\pi r^2 \exp(qV/kT_e),\tag{1.14}$$

so that

$$\bar{v}_s/\bar{v}_e = \exp(qV/kT_e),$$

or, remembering that V is negative,

$$V = -\frac{kT_e}{q}\ln\frac{\bar{v}_e}{\bar{v}_s}. \tag{1.15}$$

If we insert the values for Boltzmann's constant and the electronic charge, $T_e = 2800$ K, $\bar{v}_e = 3\times10^5$ m s^{-1}, and $\bar{v}_s = 8\times10^3$ m s^{-1}, we obtain $V = -0.63$ V.

Of course, not all spacecraft move at 8 km s^{-1}—the velocity of a near-Earth satellite. Vertical sounding rockets typically have velocities of 1 km s^{-1} or 2 km s^{-1} in the ionosphere, and their shape is far from spherical.

Because of its symmetry the flux of ions swept up by a hypersonic sphere is independent of its direction of motion, but the flux swept up by some other shape, a long cylinder, for example, depends strongly on its orientation, so that if it is moving fast and undergoing a pitching or precessing motion its potential must be continually changing.

On the other hand, for a slow-moving vehicle the flux of ions, like that of electrons, is dependent on its total surface, not the projected area in the direction of motion, and will be given by an equation like (1.12) with the subscript e changed to +. At intermediate speeds (that is, slightly sub- or supersonic)

SPACE PHYSICS
The Study of Plasmas in Space

THE universe is composed almost wholly of plasmas; the solar wind and corona, the material expanding into space from supernova explosions, and the regions around neutron stars and black holes are a few examples. With SPACE PHYSICS, THE STUDY OF PLASMAS IN SPACE (part of the distinguished Oxford Physics Series) we have a brilliant introduction to the nature and properties of these plasmas and the methods used to study them.

Written by R. L. F. Boyd, a pioneer in the use of satellite probes, SPACE PHYSICS examines plasmas starting from the earth's upper atmosphere outwards. Boyd begins by dealing with "ionospheres," including their discovery and properties, then moves to "magnetospheres" and the discovery of the earth's radiation belts by the first American satellite that was launched. Leaving the environment of the earth, Boyd considers the sun, upon whose behavior our environment largely depends. He discusses this interaction thoroughly with a look at solar activity, an ultraviolet solar spectrograph, the energy of a solar flare, and more. From the sun, he moves out to cover the galaxy and beyond with an examination of interstellar shock waves from remote supernova explosions, rapid regular and secular variations of X-ray fluxes from neutron stars (with a density like that of nuclei), and the departing footprint of "black holes."

In a subject like this, the experimental methods used to obtain raw data are almost as exciting as the data itself. Throughout SPACE PHYSICS, Boyd is careful to balance out the text between theory and experiment; he manages to make the reader not only aware of the highly sophisticated instrumentation used but also the difficulty in obtaining significant data and analyzing it in a reliable way.

Containing 56 figures, diagrams and photographs, SPACE PHYSICS is a well-illustrated, well-written introduction to the study of plasmas in space. It is a sound armchair voyage into the matter between the solid systems; matter which comprises the greater part of space.

CONTENTS

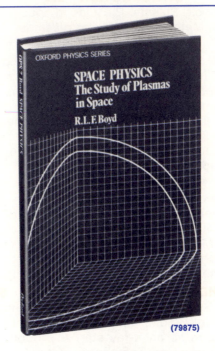

(79875)

Alternate Selection

Space Physics
The Study of Plasmas in Space

By R. L. F. Boyd

Publisher's Price $11.25

Member's Price $9.35

the situation is more complex, since both ion random and spacecraft velocities will be relevant. The behaviour for a sphere is illustrated in Fig. 1.6.

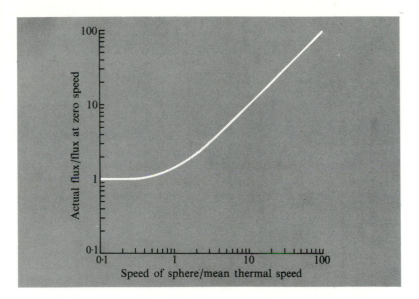

FIG. 1.6. Effect of the velocity of a sphere on the flux of particles encountered. Particles are assumed to have a thermal, that is, a Maxwellian, velocity distribution.

It is now important to inquire whether the flux of ions is not altered by the spacecraft assuming a negative potential. We might expect the field from the craft to pull in ions from a large surrounding volume. In fact this can happen, and whether it does is all a matter of scale. A negative electrode pushing away electrons and attracting ions surrounds itself with a cloud of un-neutralized ions, ever moving towards it and being collected and ever-replenished from the plasma. This space-charge cloud or sheath screens off and terminates the field from the electrode. The thickness may be calculated by a modification of the theory of the current in a space-charge limited diode, the modification taking account of the presence of some electrons.

The thickness of the sheath on a plane electrode is given fairly well by $d \sim 1\cdot3\eta^{\frac{3}{4}}h$. Here η measures the potential difference across the sheath between the electrode and surrounding space in terms of the thermal energy of the electrons. It is given by $\eta = qV/kT_e = 11\,600V/T_e$, while h is a parameter greatly used in plasma theory, known as the Debye length. It is a characteristic length of a plasma and comes from the theory of strong electrolytes: $h = (\varepsilon kT_e/n_e q^2)^{\frac{1}{2}}$.

Typical approximate values of h as a function of ionospheric altitude are as follows

height (km)	75	100–400	800	3000	13 000
Debye length, h (cm)	10	0·4	1	6	70.

Over the height range 75–3000 km it is clear that the sheath around most spacecraft will be small compared with the craft dimensions, so that it is reasonable to suppose the ions are collected over a surface equal to the projected area for a fast craft or the total surface area for a slow one.

Out in the magnetosphere, at 13 000 km, this assumption no longer holds. Moreover, as the plasma density decreases the relative importance of photo-electric emission increases. So far we have ignored the effect of this on the equilibrium potential of the spacecraft, but in the magnetosphere it may well be the controlling influence and drive the vehicle potential positive until it can retain the fairly high-energy photo-electrons produced by the intense solar ultraviolet flux.

The Langmuir probe on a spacecraft

Late in the nineteenth century Crookes introduced little probing electrodes into the plasma of the electrical glow discharge in an effort to explore the electric fields there, but it was not until the 1920s that Langmuir showed that a study of the i–V characteristic for such a probe could give three important quantities—space potential, electron–ion concentration, and electron temperature.

The spherical satellite we thought of earlier is rather like one of Crookes' little 'idle poles', as he called them; it is an isolated spherical electrode in a plasma, and it takes up a potential given by eqn (1.15). For a stationary probe we would replace $\bar{v}_s$ by $\bar{v}_+$, the mean thermal velocity of the ions, and we could then express the v's in terms of the temperatures and masses of the particles, to obtain

$$V = -\frac{kT_e}{q}\ln\left(\frac{T_e}{T_+}\frac{m_+}{m_e}\right)^{\frac{1}{2}}. \tag{1.16}$$

V here is the potential of the spacecraft or probing electrode when drawing no current or, to put it another way, it is the *floating potential* of the probe taking the zero of potential as that of the surrounding plasma. We can write

$$V = V_F - V_S,$$

where V_F stands for the floating potential of the probe and V_S for the space potential. If, as just suggested, we arbitrarily set $V_S = 0$, we have $V = V_F$. If instead we decide to measure all potentials with respect to that of the spacecraft taken as zero we have $V_S = -V$ for the potential of space.

Let us now imagine our sphere sticking out on a stalk from an even larger spacecraft whose potential is $-V_S$ with respect to space, and suppose we can vary the potential of our probing sphere with respect to the parent spacecraft. We will call this potential difference V_P for probe voltage. The potential of the probing sphere with respect to space will be

$$V = V_P - V_S, \tag{1.17}$$

and the current to it from the plasma will be given by eqn (1.14).

$$i = \tfrac{1}{4}qn_+ A\{v_+ - v_e \exp(qV/kT_e)\}. \tag{1.18}$$

This curve is conventionally plotted with positive ion currents shown negative and looks like Fig. 1.7. In practice, the dependence of sheath thickness on

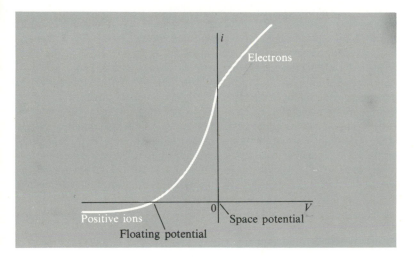

FIG. 1.7. Langmuir probe characteristic.

probe voltage usually results in poor saturation of the positive and negative branches of the curve. Nevertheless, since in practice the electron branch is about two orders of magnitude greater than the positive ion branch we can estimate the first term in eqn (1.18) reasonably accurately from the latter, and by subtraction isolate the electron current component (the second term). Then plotting ln i against V we get a straight line whose slope is q/kT_e, from which the electron temperature may be obtained. The logarithmic curve shows clearly, as a departure from linearity, the break away from the exponential when space potential is reached. It thus determines this for us and gives a value of the random electron current. Inserting this, together with the value of T_e from the slope of the curve to give $\bar{v}_e$, into eqn (1.12) enables us to solve for the plasma density n_e.

In practice, both the above theoretical and experimental procedures are substantial oversimplifications, but apart from one significant factor still to be considered they represent, in principle, the most important way in which electron temperature in the ionospheric plasma can be measured and the vehicle potential (or its reverse, the plasma potential) obtained. Important studies of the density n_e of the plasma have been made also in this way, though on the whole the accuracy here does not compare well with that obtainable by radio-frequency methods.

The remaining significant factor referred to above concerns the return to the plasma of the current drawn. So far we have completely ignored the fact that the saturation current of electrons to the (spherical) probe can only be drawn if the parent vehicle potential changes sufficiently to enable it to take an equal current of the opposite sign. Many ways of dealing with this so-called 'current-dumping' problem have been tried. By far the commonest is to arrange the probing electrode to be so small (say 10^{-4} of the area) compared with the spacecraft that a trivial change of the latter's potential restores the equilibrium. This common solution, however, means that probe dimensions are usually of the order of millimetres or centimetres rather than metres, which makes the effect of sheath size an important consideration. In fact, some analyses aimed at finding the plasma density involve a study of the form of the 'saturation' branches, which depends on the orbital motion of charges in the sheath when this is large compared to the probe radius rather than a measure of space potential and the random current at that point.

It is of interest that both the first sounding rocket and first satellite, the famous Sputnik III, to use Langmuir's technique ran into trouble because of inadequate current-dumping arrangements.

An electron temperature probe

The system originally developed for the satellite Ariel I and since employed on several other satellites makes use of the fact that the electron temperature may be found from any part on the Langmuir characteristic exponential by taking the ratio of two derivatives. Thus, if we differentiate eqn (1.18) with respect to V we can neglect the derivative of the first term and obtain

$$i'_e = \frac{q}{kT_e} i_e, \tag{1.19}$$

similarly the second derivative is

$$i''_e = \left(\frac{q}{kT_e}\right)^2 i_e, \tag{1.20}$$

so that

$$i'_e/i''_e = kT_e/q. \tag{1.21}$$

It is possible to obtain this ratio in a number of ways. The method employed on Ariel I was to use the slope and curvature of the characteristic to amplify and mix two small a.c. voltages applied to the probe. A slow sawtooth waveform was applied at the same time to ensure that, for a useful period, the curve was sampled not far negative of space potential. If we represent the a.c. voltages by $V_1 \cos(\omega_1 t + \varepsilon_1)$ and $V_2 \cos(\omega_2 t + \varepsilon_2)$, where ω_1 is small compared with ω_2, we may expand the resulting current in the form

$$i = i_e + i'_e\{V_1 \cos(\omega_1 t + \varepsilon_1) + V_2 \cos(\omega_2 t + \varepsilon_2)\}$$
$$+ i''_e\{V_1 \cos(\omega_1 t + \varepsilon_1) + V_2 \cos(\omega_2 t + \varepsilon_2)\}^2 + \text{etc.} \quad (1.22)$$

The amplitude of the component at angular frequency ω_2 is $V_2 i'_e$. This was obtained by the circuit of Fig. 1.8, whose operation is illustrated by Fig. 1.9. An automatic gain control keeps the output of the first amplifier constant, and the value of the gain control voltage provides a measure of i'_e.

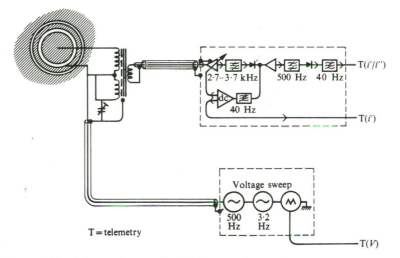

FIG. 1.8. Circuit first used on satellite Ariel 1 to obtain the temperature of ionospheric electrons by extracting the ratio of the first and second derivatives of a Langmuir probe curve (Bowen, Boyd, Henderson, and Willmore 1964).

The third term in the expansion contains the cross-modulation component $V_1 V_2 \cos(\omega_1 t + \varepsilon_1) \cos(\omega_2 t + \varepsilon_2) . i''_e$. Rectification, amplification, and filtering of this signal gives the modulation wave at frequency ω_1, which on further rectification and filtering gives a voltage to be fed to the telemeter as a measure of i''_e. Actually, because the automatic gain control keeps the carrier amplitude before rectification constant, the output to the telemeter is a measure of i'/i'', that is, of q/kT_e.

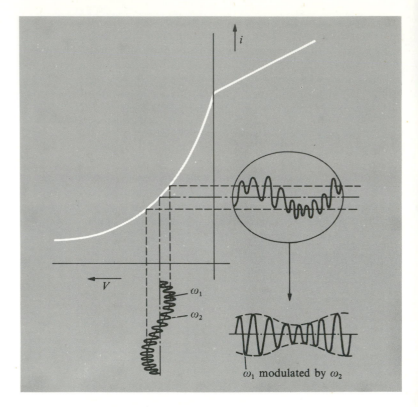

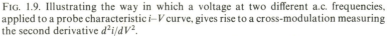

FIG. 1.9. Illustrating the way in which a voltage at two different a.c. frequencies, applied to a probe characteristic $i-V$ curve, gives rise to a cross-modulation measuring the second derivative d^2i/dV^2.

The small condenser at the end of the line feeding the probe neutralizes an unwanted component of current due to the probe capacitance. This apparatus operated on Ariel I for over 2 years and made tens of thousands of measurements. Fig. 1.10 shows an example of the variation of temperature with latitude obtained in this way. The interesting feature is the rise of temperature as we go away from the equator. The reason for this becomes clear when we recall that the equilibrium temperature of anything depends on a balance between heat input and heat output.

Electrons in the ionosphere lose their energy almost entirely in collisions with positive ions, which, because of the Coulomb law of attraction, offer a much larger collision cross-section than the neutral particles. The ions then pass the energy on to the far more numerous neutral particles, with which their

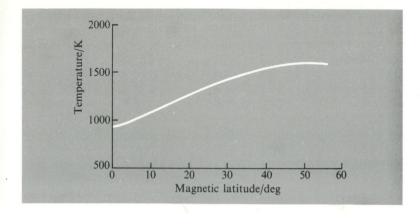

FIG. 1.10. Electron temperature in the ionosphere at an altitude of 400 km and at local noon as determined by Ariel I (Bowen, Boyd, Henderson, and Willmore 1964).

energy exchange is much more efficient than is the electrons', because of the nearer parity of mass. As we go away from the tropics the solar zenith angle χ increases, so the flux of heating and ionizing radiation from the Sun falls. For a variety of reasons, including the lower temperature of the neutral atmosphere, the ion concentration high above the F layer maximum falls with latitude more rapidly than the ionizing and heating flux, with the result that the cooling rate of the electrons drops less rapidly than the heat input, resulting in a rise of electron temperature with latitude.

The mass spectrum and temperature of ionospheric ions

In the last section we saw how the measurement of electron current to a probe in the ionospheric plasma could provide information on the temperature and concentration of the electrons. We saw also that some simple electronic processing (in an analogue rather than a digital manner) could prepare the data to reduce the demands on the information-carrying capacity of the radio telemetry link. This on-board data analysis, to extract a quantity of interest, can be very important both in simplifying the radio link and in reducing the amount of data to be recorded and stored on magnetic tape on the ground.

In this section we shall see how a virtue is made of the hypersonic velocity of a satellite, enabling us to gather information on the mass spectrum of the ions and their temperature. We have noticed already that ion and electron temperature need not be the same; the amount by which the electron temperature exceeds the ion temperature, together with information on the actual electron temperature and concentration, can provide us with a figure for the flow of solar energy into the electron gas.

The ion masses and temperature can be obtained by measuring their energy distribution function $f(E)\,dE$ which expresses the number of ions per unit volume which in the frame of reference of the satellite have an energy between E and $E+dE$.

The principal ions at satellite altitudes (say above 350 km) are H^+, H_e^+, and O^+, and they will arrive at the satellite with mean energies proportional to their mass, so that the energy spectrum may be expected to look something like Fig. 1.11. It is easy to see that this energy spectrum is also a mass spectrum. In the last section we considered electrons with a thermal, that is, a Maxwellian, energy distribution. Here the thermal velocities are superimposed on the satellite's own velocity, which is no longer negligible.

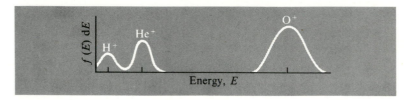

FIG. 1.11. Theoretical energy spectrum, in the frame of reference of a near-Earth satellite, of ionospheric ions of hydrogen helium and oxygen.

If we carry out the necessary algebra we find that the energy distribution for a species of ion with mass m_+, temperature T_+, and concentration n_+ has approximately the form

$$f(E)\,dE = \frac{n_+}{(2\pi m_+ kT_+)^{\frac{1}{2}}v_s}\exp-\left\{\frac{(E-\frac{1}{2}m_+v^2)^2}{2kT_+m_+v_s^2}\right\}dE, \qquad (1.23)$$

the approximation becoming more exact as the thermal energy kT_+ is smaller than the energy $\frac{1}{2}m_+v_s^2$.

This expression, though it may look a little fearsome, is simply the familiar bell-shaped 'error function', known as a Gaussian curve centred on the energy $\frac{1}{2}m_+v_s^2$. Its half-width at half height $2(kT_+\cdot\frac{1}{2}m_+v_s^2)^{\frac{1}{2}}$ therefore provides a measure of the ion temperature T_+, knowing v_s once m_+ has been found from the position of the peak.

Now if we look a little closer at the expression $2(kT_+\cdot\frac{1}{2}m_+v_s^2)^{\frac{1}{2}}$ we notice an interesting and rather surprising thing. It is twice the geometric mean of the ion thermal energy kT_+ and the mean energy in the frame of the satellite $\frac{1}{2}m_+v_s^2$, and since the latter is much greater than the former, the peak is much wider than the value kT_+ which we might have guessed as giving its order of magnitude. This unexpected result increases the accuracy with which we can measure T_+ and comes about because the satellite adds its velocity to those of the ions, the resultant being squared to obtain the energy. We can see what happens in the simple case of an ion approaching the satellite along the latter's path with an energy kT_+.

The ion velocity would be $v_+ = (2kT_+/m_+)^{\frac{1}{2}}$, so that its velocity relative to the satellite would be $v_s + v_+$ and its energy in the satellite frame

$$E = \tfrac{1}{2}m_+(v_s + v_+)^2 = \tfrac{1}{2}m_+(v_s^2 + v_+^2 + 2v_+v_s), \tag{1.24}$$

but we can neglect v_+^2 compared with v_s^2, so that the effect of the thermal velocity v_+ is to increase the ion energy in the frame of reference of the satellite by $m_+ v_+ v_s$ or, in terms of T_+, by

$$m_+\left(\frac{2kT_+}{m_+}v_s^2\right)^{\frac{1}{2}} = 2(kT_+ \cdot \tfrac{1}{2}m_+ v_s^2)^{\frac{1}{2}}. \tag{1.25}$$

We consider now how this energy spectrum of the ions may be measured. The energy distribution of ions arriving at a probe having a retarding potential V with respect to space was shown, many years ago, to be given by

$$f(E)\,dE = \frac{2}{Aq}\left(\frac{2m_+}{q}\right)^{\frac{1}{2}}V^{\frac{1}{2}}\frac{d^2i}{dV^2}\,dE. \tag{1.26}$$

The principles are the same as those of Langmuir's analysis of probe operation, except that a generalized energy distribution has been taken in place of that of Maxwell. Providing the velocity distribution is isotropic, this result is true for almost any shape of probe, for example, a cylinder or plate. Fortunately, providing the probe is spherical, it still holds for an anisotropic velocity distribution like that seen by the moving satellite.

It would appear therefore that we can obtain the energy distribution, and so the mass spectrum and temperature of ionospheric ions, by doubly differentiating a retarding potential probe curve, and we have already seen that double differentiations may be carried out by the electronic dodge of using the characteristic to mix (that is, cross-modulate) two frequencies. (Actually we could, and sometimes do, use other properties of the curve to get the second derivative, such as demodulation—that is, rectification—or harmonic generation.)

However, there remains one further instrumental problem. If we take a sphere large enough to collect sufficient ion current, in practice 10 cm or 20 cm diameter is used, and put a retarding, that is, a positive, potential on it, it will take a huge electron current and will drive the whole satellite negative because of the current-dumping problem. To overcome this we surround the probe with a concentric spherical grid at a sufficiently negative potential to reject all the electrons and prevent them reaching the probe. The interposition of the potential valley due to the grid has negligible effect on the energy distribution of the particles, just as the ultimate energy of a particle sliding on a smooth surface is unchanged by crossing a valley in its path. The small but negligible effect the grid does have is due to the fact that its

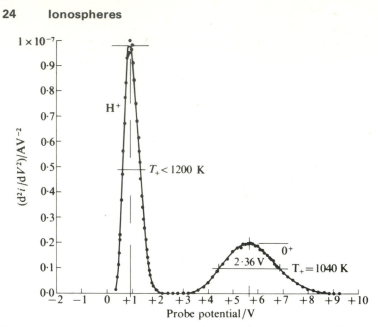

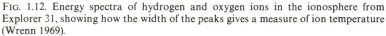

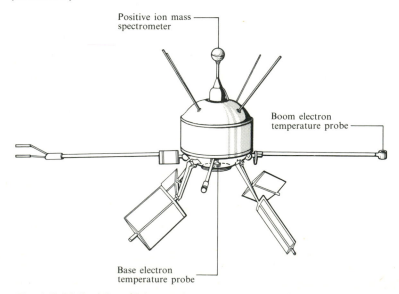

FIG. 1.12. Energy spectra of hydrogen and oxygen ions in the ionosphere from Explorer 31, showing how the width of the peaks gives a measure of ion temperature (Wrenn 1969).

FIG. 1.13. Mullard Space Science Laboratory's ionospheric instruments on Ariel I.

transparency is finite and depends on the fields around the holes in it and to the fact that these fields can change the direction of motion of a particle passing close to the edge of a hole. Fig. 1.12 shows some actual measurements of ion energy spectra made with an instrument of this kind on Explorer XXXI. Fig. 1.13 is a drawing of Ariel I carrying instrumentation to obtain the ion and electron temperature and concentration in the ionosphere between about 400 km and 1000 km. This was the first satellite to carry this kind of instrumentation. The most recent, ESRO-4, built in the U.K. was launched on a Scout four-stage solid-fuel rocket from the Californian test range in November 1972. It moves in an orbit over the poles, as did two earlier satellites in the series, carrying somewhat smaller spheres amongst their instruments. Its purpose is to study the polar regions, where energy comes in from fast particles moving on the Earth's magnetic field lines.

Distribution of ions in the topside ionosphere

We mentioned earlier in this chapter, that it is often necessary to consider the equation of hydrostatic equilibrium (1.2) as applying separately to each of the atmospheric constituents. When the constituents are ionized the situation is complicated by the presence of a small vertical electric field acting in an upward direction. It can be thought of as modifying the gravitational force mg acting on each particle to a value $mg - qE$. The field arises because of the high mobility of the electrons, which would otherwise escape from the Earth in greater numbers than the positive ions. This so-called ambipolar field can have a significant effect on the way the composition of the atmosphere varies with height. It may be calculated in the following way. The condition of quasi-neutrality over regions large compared with the Debye length requires that

$$\sum_i q_i n_i + q_e n_e = 0 \qquad (1.27)$$

where the qs and ns are respectively the charges carried and the concentrations of the various species of ion. i refers to the ith ion species and e to the electrons. The ions are in equilibrium at an altitude z if the net force on them is zero

$$0 = (qE - \bar{m}_+ g) \sum_i n_i - \frac{\mathrm{d}}{\mathrm{d}z} \sum p_i. \qquad (1.28)$$

where m_+ is the mean mass of the ions, E is the vertical electric field, and p_i is the partial pressure of the ith ion species. If we disregard the gravitational force on the electrons compared with that on the ions, and subtract eqn (1.28) from the equivalent equation for the electrons, we obtain

$$0 = \frac{\mathrm{d}p_e}{\mathrm{d}z} + \frac{\mathrm{d}\sum_i p_i}{\mathrm{d}z} + \bar{m}_+ g \sum_i n_i. \qquad (1.29)$$

There are ionospheres, notably that of the Sun, where there are large concentrations of multiply-charged ions and the dependence of the electric forces on the ionic charge plays an important part in determining the distribution of composition. In the ionospheres of the Earth and of the planets multiply-charged ions may be expected to be very minor constituents, and so we can simplify the algebra by setting $q_i = q_j$, etc. $= -q_e = q$. Moreover, the various ionic species are rather tightly coupled, as regards temperature, by their Coulomb interactions and the fact that their masses are comparable, so that for simplicity we can take $T_i = T_j$, etc., although commonly the small mass of the electrons makes $T_e > T_i$. Expanding the partial pressure derivatives as derivatives of concentration and temperature and eliminating n_e results in the following expression for the ambipolar electric field,

$$qE = \left(\frac{T_e}{T_e + T_i}\right)\bar{m}_+ g + k\left(\frac{\mathrm{d}T_i}{\mathrm{d}z} - \frac{T_i}{T_e}\cdot\frac{\mathrm{d}T_e}{\mathrm{d}z}\right). \tag{1.30}$$

We notice that, if the atmosphere is isothermal with $T_e = T_i$ $qE = \frac{1}{2}\bar{m}_+ g$, so that light ions, with mass less than $\frac{1}{2}\bar{m}_+$, will experience a net upward force and so increase in concentration upwards over a limited range (until $\frac{1}{2}\bar{m}_+$ falls to a value less than their mass). For an isothermal atmosphere, moreover, the scale height H_i of an ionized component is given in terms of the scale height H_n of the corresponding neutral component at the same temperature by

$$H_i = H_n\left(\frac{\bar{m}_+}{m_i}\;\frac{T_e}{T_e + T_i}\right), \tag{1.31}$$

which for $T_e = T_i$ gives $H_i = 2H_n$.

This shows that the scale height for the electrons in production–loss equilibrium in a Chapman layer (see p. 6) is the same as the hydrostatic equilibrium in a single-component atmosphere. In both cases it is twice the scale height of the corresponding neutral atoms.

Because of the greater scale height for helium He^+ ions than for the heavier oxygen O^+ ions and the even greater scale height for hydrogen ions (protons) H^+, the composition of the topside ionosphere moves progressively from O^+ to He^+ to H^+ as the major ion. Studies made with an ion-energy spectrometer on Ariel I show clearly how the relative concentration of He^+ and O^+ varies with latitude and time of day. When the atmosphere is hot and inflated at mid-day the O^+ ions reach to greater heights, as Fig. 1.14 shows.

Some other ionospheric studies

It would be inappropriate to end this chapter on the ionosphere as studied by spacecraft techniques without emphasizing the great areas of space technology which have contributed to our present picture of the ionosphere and to which no reference has been made here. Amongst the most important studies has been the measurement of the magnetic fields due to those currents which lead Balfour Stewart to his penetrating predictions. The study of particles with more than thermal energy is important in the auroral zones,

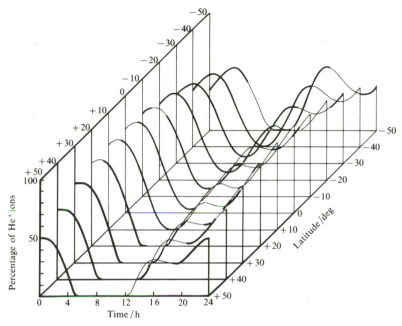

Fig. 1.14. Percentage of helium ions at 500 km in the ionosphere showing variation with latitude and time of day. Data from satellite Ariel I (Boyd and Laflin 1968).

where they give rise to the auroral light and, at times of intense solar activity, to a black-out of radio communication over the polar caps, as a result of ionization produced below 100 km. Techniques for study of these supra-thermal particles will be discussed in the next chapter, since it is in the magneto-sphere that most of these particles originate, but the study of the auroral light itself and of the photochemically generated airglow at all latitudes must be passed over with only the mention that accurate measurement of the altitude distribution of these fluorescences is an important subject, whose study with vertical sounding rockets is continuing.

The composition of the upper atmosphere and the absorption of energetic solar photons in dissociation of molecules and production of ionization are further important topics for rocket investigation. The instruments used are similar to some of those discussed in Chapter 3 though, of course, emphasis must be on fast response if the variation of solar radiation with height is to be studied.

The whole subject of the dynamics of the ionized medium is important and still inadequately investigated. Here, the interaction between tidal and thermal winds in the neutral atmosphere and the ionized conducting medium, constrained to move mainly along the lines of geomagnetism, is relevant.

2. Magnetospheres

Discovery of the Earth's radiation belts

THE winter of 1957–8 was a time of intensive activity for space research in the U.S.A. It was during the International Geophysical Year—a period of maximum in the cycle of sunspot activity and one selected especially for a co-ordinated attack by scientists from many nations on many problems in geophysics and solar terrestrial relations. A year or so earlier the Americans had announced their intention to orbit a number of small (~ 10 kg) satellites devoted to pure scientific research, to be launched by a specially-developed multi-stage rocket, Vanguard, based on their Navy's successful Viking high-altitude research vehicle. They were forestalled in the autumn however, in the start of what became in the public's eyes the 'space race', by the news that the U.S.S.R. had launched a much heavier satellite and followed it a few days later with a still larger one with a small dog incarcerated therein. These satellites, large enough, low enough, and in a sufficiently inclined orbit for almost the whole population of the world to be able to see, moved rapidly amongst the stars from east to west, scattering the rays of the Sun in the evening or morning sky.

The experiences of the dog, of course, had nothing to do with geophysics, but observations of the radio transmissions from the first two artificial satellites of our Earth, which were on a sufficiently low frequency (20 MHz and 40 MHz) to be significantly refracted and dispersed by the ionospheric plasma, did contribute to the study of the ionosphere and Sputnik III, which soon followed, carried a large weight of perhaps not very sophisticated geophysical instrumentation. Meanwhile, the Vanguard project suffered setback and failure.

In these months Professor James Van Allen, who had been doing good work on cosmic and auroral particles using, amongst other things, small rockets launched from balloons, put together a tiny payload for Explorer I, which was successfully orbited on 31 January 1958 by a Jupiter rocket—a von Braun development from his own German V-2, as the British called the Second World War weapon. There is a certain irony, perhaps even poetic justice, that this tiny satellite Explorer I made one of the most significant discoveries of the space age.

Van Allen's payload consisted of simple Geiger counters to study the flux of primary cosmic rays; at ground level the cosmic radiation is secondary, having been produced in nuclear disintegrations induced at higher altitude by impact of the primaries. At satellite altitudes, which are rarely much below 300 km, the primary flux was expected to show only the modulation with geomagnetic latitude, which is well known from observations of the secondaries and which Stömer had shown long before to be due to the deflection of the primaries towards the magnetic poles. In the event, Van Allen's counters showed a count rising to an unexpected crescendo and then, at greater height, falling to zero. The zero was due to a flux of energetic particles so greatly above the dynamic range of the counters that they were saturating, so large, in fact, that it could not possibly be due to cosmic rays but must be a geophysically-trapped population of fast charged particles.

This idea was not wholly new. Stömer himself had calculated the orbits of particles trapped in a dipole magnetic field, and Singer had suggested before their discovery that zones of energetic trapped radiation might exist. Preparations to test the idea by injecting particles from a nuclear explosion, and, of course, to study the possible global effect of such an event were being made by the U.S. Department of Defence. Experiments of this kind were carried out in 1958 using three 1 Kton yield bombs and in 1962 with a 1·4 Mton U.S. bomb and several U.S.S.R. bombs. Happily, there is now international agreement against such experiments.

Fig. 2.1 shows Van Allen's data. At perigee the counters detected a flux attributable to the primary radiation, but as apogee was approached trapped

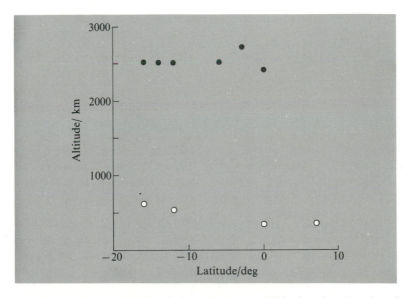

Fig. 2.1. Discovery of trapped radiation by Explorer 1. White dots show detection of a typical cosmic ray background of about 30 counts per second. Black dots show registration of no counts during 2 min intervals owing to saturation of detector by geometrically trapped particles. (After Van Allen, Ludwig, Ray, and McIlwain 1961.)

radiations, presumed to be electrons or protons, saturated the counters. Further measurements with Explorer III and Sputnik III and later satellites confirmed the existence of roughly toroidally-shaped zones of radiation encircling the Earth, with a symmetry dictated by the geomagnetic field. Even with a shielding of $10^{-4} g\, m^{-2}$ of lead over the counter windows rates of $10^8\, s^{-1}\, m^{-2}$ were reached in two 'belts', centred on equatorial distances of about 4000 km and 24 000 km. It is now known that the outer region consists mostly of electrons with energies around 1 MeV or greater, while the inner zone owes its penetrating radiation to protons mostly in the 10–100 MeV range. There is no doubt that the decay of neutrons produced lower down in the cosmic-ray interaction in the atmosphere is responsible for providing part of the proton trapped radiation, but it seems likely that the Sun is a significant source of protons and electrons in both of the zones.

The geomagnetic trapping of energetic particles

A first approximation to the Earth's magnetic field is that of a dipole (see Fig. 2.2). At any point in the field therefore, except in the plane of the equator, which is a plane of symmetry, the field has a finite gradient in the direction of the field lines. It also has a gradient in a direction normal to the field lines in a plane containing the centre of the dipole, but has no gradient in the direction orthogonal to these two—normal to the page in the diagram. Now in a field of this kind an energetic particle can be subject to three kinds of motion. The first and major motion is a quasi-helical motion along the lines of force, with the helix getting smaller in diameter and closer in pitch as the particle moves

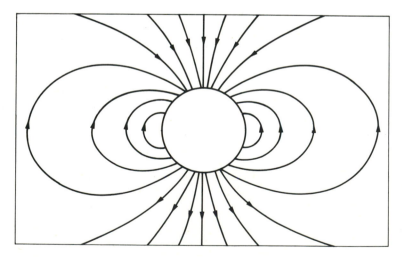

FIG. 2.2. A dipole field around a concentric sphere. Compare with Fig. 2.5.

into the stronger field region near the poles. Here, if it has not penetrated so deeply into the atmosphere as to be lost by scattering collisions, its motion reverses and it returns towards the equatorial plane and eventually to another *mirror point* near the other pole. This behaviour is a result of the field gradient along the lines of force. A second motion, due to the gradient in the direction of the Earth, causes a particle mirroring near the poles to drift eastward for electrons and westward for protons. A third effect, due to the centrifugal force on the particle in its poleward trip, shows itself as a similar drift.

All of these motions can be conveniently treated by analysing the particle's movements into (1) a circle about the lines of magnetic force, (2) a component along the lines of force, (3) a deformation of the circle due to the variation of field strength from one edge of the orbit to the other, and (4) the effect of centripetally accelerating the overall path towards the Earth, due to the general curvature of the field lines. We start, therefore, with the idea of a magnetic mirror as illustrated in Fig. 2.3, and in what follows will use a non-relativistic treatment which, though inadequate to the real situation, is nevertheless sufficiently illustrative.

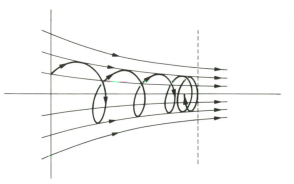

FIG. 2.3. Reflection of charged particles trapped in a converging magnetic field. A magnetic mirror.

In a uniform magnetic field a particle of charge q and mass m having a velocity $v_\perp$ normal to the lines of force moves in a circle whose radius is found simply by equating the mass times its centripetal acceleration to the magnetic force

$$\left. \begin{array}{c} mv_\perp^2/r = qv_\perp B_z \\[2mm] r = mv_\perp/qB_z. \end{array} \right\} \tag{2.1}$$

v/r is called the *angular gyrofrequency*.

We note that the radius is proportional to the particle momentum and inversely proportional to the magnetic induction.

Now if the particle has a component of velocity $v_\parallel$ along the line of force the circle becomes stretched out into a helix whose *pitch angle* Ψ is defined by

$$\tan \Psi = v_\perp/v_\parallel. \tag{2.2}$$

If the field has a gradient in the z direction and the particle is moving into the higher field, the helix pitch becomes less and less as the field increases. This motion can be analysed for the geomagnetic situation, where the change in field occurring during a single turn of the helix is negligible for the vast majority of particles, by considering the circulating particle as a little dipole magnet in a field gradient.

The circulating current due to one particle is

$$i = qv_\perp/2\pi r \tag{2.3}$$

and its magnetic moment is

$$M = \pi r^2 i = \tfrac{1}{2}qv_\perp r \tag{2.4}$$

Substituting r from eqn (2.1) gives

$$M = \tfrac{1}{2}mv_\perp^2/B_z = K_\perp/B_z, \tag{2.5}$$

where $K_\perp$ is the energy associated with the gyro motion.

Now the force on a magnetic moment M aligned parallel to a field gradient $\mathrm{d}B_z/\mathrm{d}z$ is

$$F_z = -M(\mathrm{d}B_z/\mathrm{d}z). \tag{2.6}$$

Equating this force to the rate of change of momentum parallel to the field lines we obtain

$$m\frac{\mathrm{d}v_\parallel}{\mathrm{d}t} = -M\frac{\mathrm{d}B_z}{\mathrm{d}z}, \tag{2.7}$$

which can be simplified by multiplying throughout by $v_\parallel = \mathrm{d}z/\mathrm{d}t$ to give

$$mv_\parallel \frac{\mathrm{d}v_\parallel}{\mathrm{d}t} = -M\frac{\mathrm{d}B_z}{\mathrm{d}z}\frac{\mathrm{d}z}{\mathrm{d}t}$$

that is,

$$\left.\begin{array}{c}\\[1em]\\[1em]\end{array}\right\} \tag{2.8}$$

$$\frac{\mathrm{d}}{\mathrm{d}t}(\tfrac{1}{2}mv_\parallel^2) = -M\frac{\mathrm{d}B_z}{\mathrm{d}t}.$$

Here $\mathrm{d}B_z/\mathrm{d}t$ is the rate of change of B_z with particle *position* (there being no change of the field itself with time).

Conservation of energy requires that the rate of change of energy associated with motion parallel to the field $\mathrm{d}(\tfrac{1}{2}mv_\parallel^2)/\mathrm{d}t$ be equal and opposite to the rate of change of that associated with the circulating motion,

$$\frac{d}{dt}(\tfrac{1}{2}mv_{\parallel}^2) = -\frac{d}{dt}(\tfrac{1}{2}mv_{\perp}^2)$$

or   (2.9)

$$\frac{dK_{\parallel}}{dt} = -\frac{dK_{\perp}}{dt},$$

where $K_{\parallel}$ is the energy associated with $v_{\parallel}$, and the total kinetic energy $K = K_{\parallel} + K_{\perp}$. Eqns (2.8) and (2.9) together give

$$\frac{dK_{\perp}}{dt} = M\frac{dB_z}{dt},\qquad(2.10)$$

which from eqn (2.5) leads to

$$\frac{dMB_z}{dt} = M\frac{dB_z}{dt}.\qquad(2.11)$$

In other words *the magnetic moment is a constant of the motion.*

The constancy of the magnetic moment of the particle as it moves along the lines of force can readily be shown to imply the constancy of two other quantities: the flux enclosed in a turn of the helix and the angular momentum about the field lines. All these quantities are called *adiabatic invariants,* implying that the invariance is an approximation depending on a *slow* variation of the field encountered by the particle compared to the gyro-frequency.

It is easy to see that eqn (2.5) implies that a value of B_z might be reached for which all the kinetic energy is associated with $K_{\perp}$. At such a point—a mirror point—the particle is reflected, since the force of eqn (2.7) continues to operate on the magnetic moment and $v_{\parallel}$ reverses.

At the mirror point, $K_{\parallel} = 0$ and, from (2.5),

$$B_z = K/M.\qquad(2.12)$$

Azimuthal drift of trapped particles

We mentioned earlier that the increase in field in the x direction dB_z/dx (see Fig. 2.2) results in a drift around the Earth transverse to the field. We can think of this as arising from the fact that the field is slightly different on each side of the helix. For an approximate treatment consider Fig. 2.4 and let us suppose that the field B_z has the constant mean value $B_z + \tfrac{1}{2}r(dB_z/dx)$ below the centre line and $B_z - \tfrac{1}{2}r(dB_z/dx)$ above the centre line.

Now the drift per revolution is $\sim 2(r_2 - r_1)$, where, from eqn (2.1),

$$r = (mv_{\perp}/q)B_z,$$

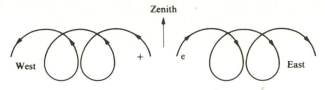

FIG. 2.4. Motion of electrons and positive ions trapped by the Earth's field as a result of the decrease in the field towards the zenith.

so

$$r_1 = \frac{mv_\perp}{q}\left(B_z + \frac{r}{2}\frac{dB_z}{dx}\right)^{-1} \tag{2.13}$$

and

$$r_2 = \frac{mv_\perp}{q}\left(B_z - \frac{r}{2}\frac{dB_z}{dx}\right)^{-1}. \tag{2.14}$$

These expressions can be simplified by expanding binomially to give

$$r_1 = \frac{mv_\perp}{q}B_z\left(1 - \frac{r}{2B_z}\frac{dB_z}{dx}\right), \tag{2.15}$$

so that

$$r_2 - r_1 = \frac{mv_\perp}{qB_z}\cdot\frac{r}{B_z}\cdot\frac{dB_z}{dx}. \tag{2.16}$$

The drift velocity $v_{d_1} \simeq \dfrac{2(r_2-r_1)v_\perp}{2\pi r}$,

so that

$$v_{d_1} \simeq \frac{m}{q}\cdot\frac{1}{\pi}\cdot\left(\frac{v_\perp}{B_z}\right)^2\cdot\frac{dB_z}{dx}. \tag{2.17}$$

A more exact treatment leads to

$$v_{d_1} = \frac{m}{q}\cdot\frac{1}{2}\cdot\left(\frac{v_\perp}{B_z}\right)^2\cdot\frac{dB_z}{dx} = \frac{1}{q}\frac{K_\perp}{B_z^2}\cdot\frac{dB_z}{dx}. \tag{2.18}$$

We notice that this drift depends on the sign of q and is independent of $v_\parallel$. The second drift effect however is due to $v_\parallel$ and arises from the fact that the motion must be modified to give rise to a force which will constrain the particle to move on the curved field lines.

If R is the radius of curvature of the lines at a given point, the required force is $mv_\parallel^2/R$. This force can be thought of as arising from the drift of the particle across the field (the Lorentz force), so that

$$qv_{d_2}B_z = mv_{\parallel}^2/R$$

or

$$v_{d_2} = mv_{\parallel}^2/qB_zR$$

(2.19)

It can be shown that

$$\frac{1}{R} = \frac{1}{B_z} \cdot \frac{dB_z}{dx},$$

so that

$$v_{d_2} = \frac{mv_{\parallel}^2}{qB_z^2} \cdot \frac{dB_z}{dx}.$$

(2.20)

We notice again that the drift depends on the sign of q but not on the sign of $v_{\parallel}$. So the more rigorous treatment using the geometry of the field to express dB_z/dx in terms of B_z and R and taking the true form of the path represented by Fig. 2.4 leads to the result that the total drift velocity

$$v_d = v_{d_1} + v_{d_2} = (\tfrac{1}{2}v_{\perp}^2 + v_{\parallel}^2)m/qB_zR = (K_{\perp} + 2K_{\parallel})\frac{dB_z}{dx} qB_z^2,$$

(2.21)

and clearly depends on the energy of the particle, although it is not independent of the way the energy is shared between components of velocity parallel and normal to the field.

Because the positive ions in the magnetosphere drift in the opposite direction to the electrons their currents add to give a *ring current*. At the end of Chapter 1 we mentioned currents in the ionosphere driven by tidal motions of the air; this ring current, however, flows much higher up. The former is principally responsible for the daily variations in the magnetic intensity and direction at the Earth's surface, while the latter is partly responsible for the variation occurring after a solar outburst and called a magnetic storm. Electrons drift eastwards and protons westward around the globe in about 30 min.

The magnetospheric plasma

The ionospheric plasma consists predominantly of protons and electrons as greater and greater heights are reached. The temperature rises to several thousand degrees, and this fact, together with the ambipolar field and the lightness of the major ion, results in a scale height (that is, altitude over which the concentration falls by $1/e$—see p. 6 and p. 26) of several thousand kilometres. Moreover, above 500 km altitude the mean free path of the atoms becomes so long that they are moving in orbits around the Earth, which may or may not penetrate into the deeper atmosphere and be scattered out by collisions—a region known as the *exosphere*. Charged particles can be scattered by the Coulomb forces of other charges, so that the thermal ions and electrons interact together to greater heights than do the neutral particles. This Coulomb interaction, however, depends strongly on the particle energies and is quite negligible for the energetic trapped radiation. Such particles are lost from the trapping zones if they dive sufficiently deep into the atmosphere before encountering their mirror point, which depends, of course, on the individual pitch angle.

We give the name *magnetosphere* to the region in which, because of the lightness of the ions, the diminution in the gravitational field, and the long free paths, the motions of the particles are primarily controlled by the Earth's magnetic field. This statement itself implies that the particles are (mainly) charged, a result which comes about because the ambipolar field doubles the scale height for the charged component (p. 26). There are two main populations of particles in the magnetosphere, the thermal plasma of the topside ionosphere, whose concentration is wholly determined by dynamical considerations, production and recombination being negligible at these heights, and the energetic trapped radiation discovered by Van Allen. The former population represents a highly conducting medium which maintains the electrical neutrality of the region and through which the energetic particles make their way. It must be recognized, however, that the thermal plasma is subject to drifting and mirroring, as are the energetic particles, though scattering perturbations of their paths are far more important.

So far we have thought of the magnetic field of the Earth as trapping the energetic radiation. In a very real sense, however, the thermal plasma of the ionosphere traps the magnetic field at magnetospheric heights. We are all familiar with the fact that a magnetic field can penetrate a good conductor only slowly, so changing the field in it. What happens is that the changing field induces currents in the good conductor which tend to cancel the field where it would penetrate and to strengthen it outside that region.

It is reasonable to enquire under what conditions a plasma controls the field and when the field moves the plasma. It is a matter of energy density; when the thermal energy density in the plasma exceeds the magnetic energy density in the field the plasma controls the field, and if there are already field lines in the plasma they will tend to be carried with it in any motion which the plasma has. Such a field is said to be *frozen in*. If the energy density in the magnetic field exceeds the thermal energy density in the plasma, the plasma will tend to be moved by any motion of the field lines.

The lines of the geomagnetic field are frozen in to the ionospheric–magnetospheric plasma, and motion of the plasma distorts the field. On the other hand, the trapped radiation is so much less dense that, in spite of the great energy of the particles, they do not significantly disturb the field as a whole, as is evidenced by the fact that geomagnetic storm variations at the Earth are a small perturbation on the mean value.

The solar wind

We shall see in the next chapter that the outer atmosphere of the Sun, the *corona*, is at a temperature of order a million degrees. At such a high temperature the more energetic electrons and ions in the corona have sufficient energy to escape from the Sun's gravity, and so boil off into interplanetary space as the *solar wind*.

At the orbit of the Earth this solar wind normally has a velocity of about $400 \, \mathrm{km \, s^{-1}}$ or a little less, a plasma density a little under $10^7 \, \mathrm{m^{-3}}$, and a temperature of order $10^5 \, \mathrm{K}$. With this high concentration of particles it is a highly conducting medium. Its existence was first postulated to account for the way comets' tails are swept back from the Sun.

Now if we think of the Earth as the source of a dipole magnetic field subjected to the blast of charged particles in the solar wind, we can see that, up to a point, the solar plasma will carry the Earth's field before it, compressing it until the energy density in it reaches that of the kinetic pressure of the wind. At this point the Earth's field will distort no further and the solar wind splits and flows around the magnetosphere.

A plasma can support various kinds of waves. There are ordinary electromagnetic waves like light and radio which, unless they are at frequencies near to or below the critical frequency, are not greatly affected by the plasma. Then there are *ion acoustic waves*, in which there is longitudinal movement of the ions accompanied by an adjustment in the concentration of electrons to maintain quasi-neutrality. At higher frequencies there are *electron acoustic waves*, in which there is longitudinal motion of the electrons with negligible motion of the ions. The resulting departure from neutrality introduces electric fields which are much greater than those arising in ion acoustic waves of the same amplitude and which have a controlling effect on the phenomenon. If a magnetic field is present, yet other waves, generally called *Alfvén waves*, are possible. One mode involves motion transverse to the direction of propagation, like waves in a stretched string, and indeed may be pictured as displacements of the magnetic field lines and their associated plasma.

The motion of the solar wind is strongly supersonic with respect to the velocities of ion acoustic (and Alfvén) waves, with the result that a shock wave is set up, standing off from the boundary of the magnetosphere and separated from it by the *magnetosheath*, a turbulent and disturbed region. Behind the Earth the lines of force, which otherwise would have radiated from the polar regions, are swept back to form a long tail of plasma-entrapped field extending at least to the orbit of the Moon (lunar satellites have detected it).

Clearly, since the lines in the tail from the north and south polar regions are of opposite sign, there is a nodal region between them. It is known as the *neutral sheet.* Here the fields cancel and their lines join up from time to time instead of running parallel, but since the magnetic energy $B^2/2\mu_0 m^{-3}$ cannot be annihilated it turns up as kinetic energy of accelerated plasma particles. This is not as surprising as it might seem at first sight when it is remembered that the field away out in the tail is due to currents of particles in the plasma. What we have, therefore, is a mechanism, by no means fully understood, for converting the ordered motion of the plasma in current flows to the disordered motion of no current with some highly accelerated particles. It is thought to be these particles flowing back along the field lines which give rise to many of the auroral phenomena.

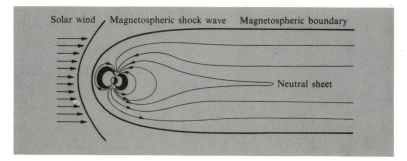

FIG. 2.5. Magnetosphere. The dark crescents represent the regions of trapped energetic particles (Van Allen radiation). The turbulent region between the shock wave and the magnetospheric boundary is known as the magnetosheath.

Fig. 2.5 is a diagram of the magnetosphere which, as can be seen, is anything but spherical. In fact the whole complex of phenomena is very involved, with electric currents, convection currents of plasma, oscillations, and disturbances, and little understood accelerating mechanisms for particles, almost certainly including electric fields. Almost all the details of the picture have been obtained from satellites or planetary probes with highly eccentric orbits threading through the region. Any study of this kind leaves an ambiguity as to whether a changing parameter as measured on the spacecraft is changing spatially, temporally, or both. In order to resolve these problems there is an increasing emphasis on using pairs of orbiting craft or geostationary craft together with sounding rockets in the polar regions and on correlated spacecraft and ground-based observations, especially of magnetic field.

The plasmapause

A glance at Fig. 2.5 shows that there are some field lines that are closed and others that are open and mingle eventually with the fields of interplanetary space, which have themselves been carried out from the Sun in the plasma of the solar wind. Closed lines connect *conjugate points* in the ionosphere of opposite hemispheres, and many correlated phenomena occur at these points. For example, photo-electrons ejected by sunlight in the summer polar region may be detected at the conjugate point of the dark winter arctic. Trapping of plasma, energetic or thermal, is only possible where the field lines are closed. Their outer boundary therefore marks the absolute limit of the trapped radiation zones. It also marks a big drop in thermal plasma density, the so-called *plasmapause* for beyond this the plasma can stream out to space.

Broadly speaking, the equatorwards limits of the auroral regions are marked by the same boundary between open and closed lines, though the main auroral activity occurs at the base of the lines which pass close to the neutral sheet.

Magnetic storms and aurorae

We saw in Chapter 1 that a connection between the Sun and the behaviour of the magnetic field of the Earth was recognized about a century ago. One of the indices of the solar influence in this respect was the presence of sunspots, which are well known to correlate with the likelihood of solar activity (see next chapter).

As might be expected from so turbulent a source as the Sun the solar wind is continually fluctuating in particle concentration and energy. At times of *solar storms*, which, as we shall see, are localized phenomena on the Sun, these changes show up as a substantial increase in the energy density of the wind and result in two main effects—distortion of the magnetospheric field and the plasma entrapped in it and the injection of high-energy particles.

The cloud of plasma ejected during a solar storm has velocity around twice that of the quiet solar wind and reaches the Earth about a couple of days after the outburst, having travelled a somewhat tortuous path owing to deflections in the interplanetary magnetic field. The first effect of the arriving plasma is to compress the magnetosphere and so to give rise to an increase in the magnetic field measured at the Earth's surface (see Fig. 2.6). A few hours later, however, the field change reverses, and the field falls by a fraction of a per cent below its normal value. This appears to be due to the replenishment of high-energy particles in the trapped radiation belts, which by their circulation around the Earth reduce the field at the surface. These particles must enter the magnetosphere along open field lines, and so probably come in from the direction of the tail. The field at the Earth reaches its minimum value about half a day after the *commencement* of the magnetic storm and recovers in the next two or three days. This series of events is called a *magnetic storm*.

Polar aurorae have been recorded for at least two millenia, and have been carefully observed for three centues. Aurora was, of course, the Goddess of Dawn and the Northern 'Dawn' was appropriately named *Aurora Borealis* by Gassendi in the seventeenth century, while the southern display was named by Cook, about a hundred years later, *Aurora Australis*. The idea of dawn

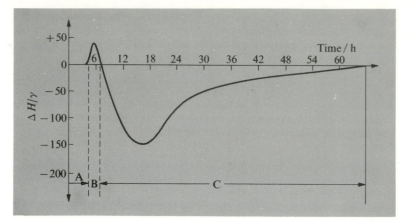

FIG. 2.6. A magnetic storm. Change in the horizontal component of the Earth's magnetic field in units of $\gamma (= 10^{-9}$ T) as a function of time. A = sudden commencement; B = initial phase—positive change; C = main phase—negative change.

refers only to the light, not the time. In fact, auroral activity is normally greatest near midnight.

Since 1722 it has been known that the *Aurora Polaris* is associated with magnetic activity, and since the study of phenomena of the electric discharge, at the end of last century, it has been clear that the light was due to the excitation of atmospheric gases by energetic charged particles, deflected by the geomagnetic field towards the polar regions. More recently, radar studies have revealed the presence of ionization. Most of the light, accompanied by ionization, is caused by electrons in the energy range around 10 keV at flux intensities of about $10^{14} \text{m}^{-2} \text{s}^{-1}$. However, protons of ten times this energy are also important, and sometimes cause so much ionization at relatively low altitude over the polar cap as to result in a *polar black-out* of radio propagation over the poles.

It would be natural to suppose that the particles giving rise to the aurorae come simply from the solar wind. It appears, however, that the true explanation is more complex and that most of the particles gain their energy as a result of electric fields or varying magnetic fields in the magnetosphere and its tail. The recoupling of lines of force in the tail probably plays a very important part here. At present these phenomena are poorly understood, and much spacecraft work being planned is aimed at studying the energy gain and exchange mechanisms in the magnetosphere and tail and the behaviour of the polar ionosphere.

Amongst the most important measurements to be made in the Earth's magnetosphere, and indeed in any planetary magnetosphere, are the spectra of energy, pitch angle, and mass of charged particles over the range of energies

from thermal to several keV, together with magnetic and electric field distributions and the intensity and frequency spectrum of electromagnetic
waves. Very energetic particles are studied by the detector techniques of
nuclear physics, Geiger counters, scintillators, and so on. Thermal particles
are studied by probe techniques, as discussed in Chapter 1, but, as the plasma
density decreases and the role of photo-electrons from the probe or spacecraft
becomes more and more dominant, measurement by retarding probes gets
very difficult, so that Langmuir probe techniques find little application beyond
the plasmapause.

Electromagnetic waves are measured by radio receivers, the major technical
problems, especially at very low frequencies, concerning the aerials. Long
aerials are deployed from satellites by unreeling a springy tape which folds
into a tube after the manner of a carpenter's rule. In the next sections we will
discuss techniques for spacecraft measurements of the spectra of suprathermal
particles and of electric and magnetic fields.

Suprathermal particle analysers

The velocity spectra of charged particles between energies of order 1 V and
100 keV or so are nowadays usually measured by curved-place electrostatic
deflectors. Thermal plasmas, as we have seen, are best studied by a retarding-
potential analyser of some sort (for example, a Langmuir probe), since spacecraft potential with respect to the surrounding plasma has an important
influence on the measurement and therefore must be determined (see Chapter
1). Other techniques—magnetic deflection methods, retardation systems with a
planar grided geometry, and semiconductor devices—have been used, but the
electrostatic deflection system which we shall describe, with its hemispherical
geometry, has the great advantage of high detection efficiency, with focusing
in two planes making it suitable for measuring the flux approaching in a
defined solid angle. The instruments under construction for incorporation in
the GEOS satellite are typical of the latest technology of this kind.

GEOS is a European Space Research Organization satellite to be placed
in geostationary orbit, a distance of 6·6 Earth radii. The satellite will study
particle energy spectra, electric and magnetic fields, and electromagnetic
waves at this altitude above the equator out in the magnetosphere. Its purpose
is to seek to elucidate the manner in which interaction of the solar wind with
the magnetosphere, especially at times of solar–magnetic disturbance, gives
rise to auroral fluxes of high-energy particles, a ring current of electrons and
protons, and excites a variety of electromagnetic waves. The two suprathermal
particle analysers on board will measure the fluxes of electrons and protons
moving in a north–south direction and the fluxes in the equatorial plane, thus
enabling pitch angle and energy distributions to be determined.

Fig. 2.7 illustrates diagrammatically the manner in which a cone of particle
rays entering between a pair of concentric hemispheres may be focused

diametrically opposite and may be analysed for energy. The field between concentric hemispheres is proportional to $1/r^2$; however, an approximate theory may take $\Delta r = r_2 - r_1 \ll r_2$, where r_1 and r_2 are the hemisphere radii, so that the field is almost constant at $(V_2 - V_1)/\Delta r$, where V_1 and V_2 are the potentials on the electrodes.

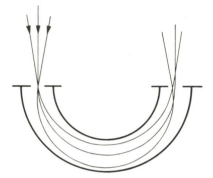

FIG. 2.7. Focusing action of concentric hemispheres with a radial electric field between them.

Then equating the electric force to the centrifugal force

$$q\Delta V/\Delta r = mv^2/r = 2qE/r, \qquad (2.22)$$

where E is the particle energy in electron volts, so that

$$E = \tfrac{1}{2}V(\Delta V/\Delta r).$$

In practice the choice of Δr is determined by the angle of the cone of rays and the range of energies accepted by the sizes of the entry and exit orifices. It is easy to see that, on the approximation $\Delta r \ll r$, the path radius is constant and so first-order focusing occurs in the plane of the diagram, since the diametral distance from entry to exit point is the constant diameter of the circle. In the plane normal to the diagram focusing arises from the spherical symmetry.

Focusing is important as it makes it possible to accept a sufficiently wide pencil of rays and still to get adequate energy resolution.

Fig. 2.8 depicts one of these analysers arranged for use with electrons. The system of plates defining the entry cone and the gauze outer hemisphere are arranged to minimize the possibility of electrons, striking the defining system, being scattered into the analyser and to reduce the interference from photoelectrons produced by sunlight striking the hemisphere surface. Protons or electrons can be studied by switching the sign of the potentials applied while the energy distribution is found by stepping the potentials on the hemispheres through the appropriate range. The actual hemisphere potentials are adjusted

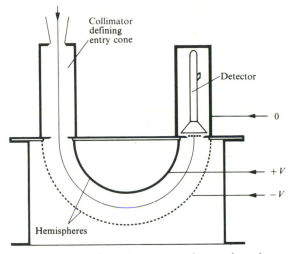

Fig. 2.8. Hemispherical electrostatic analyser arranged to analyse electrons and detect them with a channel electron multiplier (see Fig. 2.9) (Mullard Space Science Laboratory).

so that the potential at the entry orifice corresponds as closely as possible to the plasma potential outside.

The particle detector in this system is a channel electron multiplier. These devices are widely used in space science for the measurement of photon or particle fluxes. Since at low energies their detection efficiency falls off rapidly, a voltage of $+500$ V for electrons or -3000 V for protons is maintained between the sensitive cathode and a high-transparency fine-mesh grid which covers the exit orifice. The mode of operation of a channel electron multiplier is illustrated in Fig. 2.9. It consists of a glass tube, which may be coiled into a spiral, and which has a resistive coating on the inside. A single photon or charged particle striking the negative end produces further electrons, which find themselves in a longitudinal field due to the high positive potential applied to the other end. These particles strike the wall after acquiring some energy in

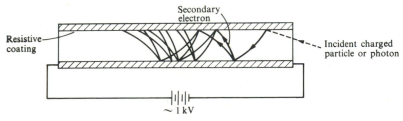

Fig. 2.9. Diagram of a channel electron multiplier.

the field and give rise to further secondaries, a process which repeats itself until a pulse of 10^4–10^5 electrons arrives at the anode end, where it can be detected as a pulse on the current conducted along the wall of the tube or may be extracted from the end by a suitable field.

Magnetic-field measurements

Magnetic fields in planetary physics are frequently measured in γ, where $10^9\ \gamma = 1$ T, that is, Wb m^{-2}. The field at the surface of the Earth is about 50 000 γ, and its strength falls roughly according to the inverse cube of the distance from the centre of the Earth's dipole. Outside the magnetosphere, however, the ambient field has a value around or below 10 γ. Spacecraft magnetometers therefore must have a wide dynamic range and constancy of calibration; more difficult still, the craft itself must have a high degree of magnetic cleanliness. A few screws made in a wrong stainless steel alloy, if included in another instrument, could render it unsuitable for flight even though their effect on that instrument itself was totally negligible. Spacecraft magnetometers may be vector or scalar instruments. Obviously the former have value apart from a study of ambient fields as a means of aspect determination of sounding rockets and near-Earth satellites when the field direction is sufficiently well known.

An early type of magnetometer which is insensitive to a constant field from the vehicle itself is a search coil on a spinning satellite. Such a device can measure the magnitude and azimuthal direction of the component of field normal to the spin axis. Its principal disadvantages are the fact that the calibration is not absolute and the low signal (even in many thousands of turns on a highly permeable core) due to the low spin rates of satellites.

A somewhat different system of three fixed orthogonal coils has been used on stabilized spacecraft. Here the array may be thought of as a very low-frequency aerial and is sensitive only to changes in the ambient field. Such changes may arise as a result of magneto-hydrodynamic (Alfvén) waves, but a serious problem exists in the difficulty of discriminating in favour of temporal against spatial changes. Increasingly, the resolution of this type of difficulty is being sought by the use of twin satellites spatially separated in the same or related orbits.

The two principal types of magnetometer used in spacecraft work are the flux gate, which is a non-absolute vector instrument, and systems depending on the Zeeman splitting of spectral lines by an ambient magnetic field. The latter instruments use an alkali vapour or helium as the optically active substance and detect the frequency corresponding to the splitting of the line by the magnetic field which is in the radio-frequency range. These devices are absolute in their calibration and very sensitive. Most versions have been scalar instruments, sometimes used in conjunction with a flux-gate array, to give directional information, though vector arrangements are possible.

The description of a Zeeman-effect magnetometer is rather difficult without a fair understanding of atomic spectra. Important as the atomic magneto-meters are, detailed description will be limited therefore to the flux-gate system.

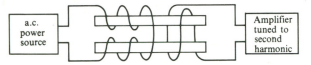

FIG. 2.10. Flux-gate magnetometer.

Fig. 2.10 illustrates the elements of a flux-gate sensor head. A pair of rods of high-permeability material is subject to a magnetic induction in the same direction as a result of the component of ambient field along their length. At the same time, they are driven around their very slender hysteresis loops in opposite directions by an alternating current in oppositely-wound primary coils. A single secondary coil around both rods measures the rate of change of net induction in the two rods. It is easy to see that when the ambient field vector in the direction of the rods is zero there is no output from the secondary. This situation is illustrated at the top of Fig. 2.11, in which the slender hysteresis

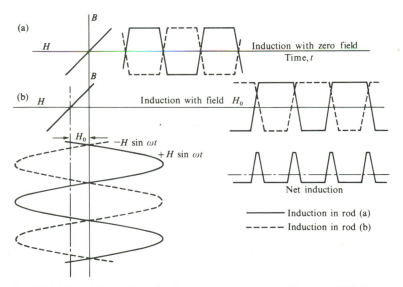

FIG. 2.11. Mode of operation of a flux-gate magnetometer. The short thick lines at 45° represent idealized B-H hysteresis loops. The a.c. field with zero bias in (a) and bias H_0 in (b) produces a net induction of zero in case (a) and a second harmonic in case (b).

loop has been represented as a thick line and in which the a.c. fields applied oppositely to each rod are shown driving the magnetic material well beyond saturation.

An ambient field H_0 is shown operating in the second part of the figure by displacing the hysteresis curve by H_0 to the left. Clearly the ambient field only affects the cores at the instants when the bias fields are very small. It is in this sense that the flux 'gate' opens and closes again as soon as saturation is reached. But the phase of opening and shutting of the flux 'gate' by the bias fields is different for the two rods, since when H_0 adds to the bias field in one rod it subtracts from it in the other. The inductions in the two rods no longer exactly cancel but instead give rise to the net induction shown. This net induction has a fundamental frequency twice that of the bias fields together with higher even harmonies. A suitable filter and amplifier arrangement selects and measures this second harmonic.

Electric-field measurements

The measurement of electric fields in space raises some important theoretical and difficult practical problems. The crux of the matter lies in the fact that electric and magnetic fields are, in essence, different manifestations of a single phenomenon and are intimately related by the fact that motion across a magnetic field gives rise to (generates, we sometimes say) an electric field. This is familiar enough. It lies at the heart of all practical conversion of mechanical to electrical energy on a large scale. The relationship is

$$E' = vB \sin \theta, \tag{2.23}$$

or in vector notation (which removes the need to apply a hand rule to determine the sign),

$$E' = v \times B, \tag{2.23'}$$

where θ is the angle between the velocity and magnetic vectors v and B, and E' is the electric field in the frame moving with velocity v relative to the frame in which B is measured.

A similar relationship exists embodying the fact that motion across an electric field gives rise to a magnetic field. This is less familiar as the effect only becomes large at relativistic velocities under which circumstances both equations require slight modifications. The relationship between E' in the moving frame of reference and E and B in the 'stationary' one is

$$B' = \mu_0 \varepsilon_0 v \sin \theta, \tag{2.24}$$

or again in vector notation

$$B' = \mu_0 \varepsilon_0 v \times E. \tag{2.24'}$$

The factor $\mu_0 \varepsilon = c^{-2}$ shows that the induction of a significant magnetic field by movement through an electric field would require very large values for the latter or for the velocity.

The theoretical problem is to decide on the appropriate frame of reference (and keep to it). We can consider the problem by looking at Fig. 2.5 and asking ourselves: should we specify the electric field at some point of interest in the magnetosphere, in a frame at rest relative to distant stars or relative to the Sun or relative to the Earth–Sun line or to the surface of the (rotating) Earth? Or if the measurement is made from a satellite is it sufficient to record simply the electric field as seen by the satellite?

In order to answer these questions we need to inquire further why the electric field is of interest. The comprehensive answer to this is: primarily because of its effect on charged particles. A more detailed inquiry would consider the field as a source of energy which may later appear as auroral excitation, ionization, electromagnetic or magneto-hydrodynamic waves, and so on. It would also consider the action of any fields in transporting charged particles and possibly in conveying, through them, momentum to neutral particles. Merely to remark this last point is to suggest a possible importance in the reverse process—the generation of electric fields by the transport of plasma through the geomagnetic field under the influence of neutral particle winds. Indeed this phenomenon is very important, it being atmospheric tidal winds, which generate the e.m.f.s responsible for driving the currents giving rise to the daily magnetic variations. So looking at the problem the other way, we see that the whole study of the origin of electric fields is important.

We saw earlier in this chapter that charged particles in the magnetosphere and upper ionosphere will be constrained to follow quasi-helical paths along the field lines. The magnetic field has only a second-order effect on motion along the lines resulting in the mirroring phenomena (second-order when comparing the mirroring distance with the radius of the helix). An electric field along the lines of force therefore acts on particles which are virtually unconstrained in its direction of action. High conductivity along the lines of force implies that any d.c. field would soon be largely neutralized by a redistribution of charge, so that fields may be expected to be small, perhaps of the order of a microvolt per metre, and the lines of force in regions where particle collision rates are small roughly delineate equipotentials. Electric fields acting normal to the magnetic field have a very different effect. The conductivity across the lines of force tends to zero as the collision frequency tends to zero, and the effect of the field is to cause charges of both sign to drift in the same direction at right angles to both electric and magnetic field, while executing their orbital (helical) motion. This behaviour is a consequence of eqn (2.23), as we shall show.

Consider an electric field E at right angles to a magnetic field B, then a frame of reference may be found travelling with a velocity v at right angles to both B and E in which the field will be

$$E' = E + vB. \tag{2.25}$$

Now if we adjust the motion of the new frame until no electric field is seen $(E' = 0)$, we have

$$v = -E/B, \tag{2.26}$$

and in this frame of reference the particle will be undergoing an unperturbed circular or helical motion according to eqn (2.1). To an observer in the original frame containing both E and B the particle will be seen to be moving with velocity

$$u(= -v) = E/B \tag{2.27}$$

superimposed on its orbital velocity. In other words, the effect of crossed electric fields on charges of both sign *in vacuo* is to cause a drift at right angles to the fields with the same velocity (and therefore no net current) for both charges, independent of sign or mass, given by eqn (2.27). Of course, any component of motion that the particles have along the magnetic field lines is left unchanged.

This effect written in vector notation is

$$u = E \times B/B^2.$$

Since this trochoidal drift is normal to the electric field it is along an equipotential surface and, in the absence of collisions, no energy is expended by the field.

As a result of the foregoing, substantial d.c. fields can only be expected in the magnetosphere across the lines of magnetic force, and their main effect is particle drift in these distant regions. However, since the lines of magnetic force are also equipotentials the systems of lines normal to the electric field define high-conductivity equipotential surfaces or shells which can conduct the potential differences down into regions of the atmosphere where collisions between particles make possible current flow from shell to shell by scattering the particles out of their trochoidal tracks. It is in this way that fields in the magnetosphere can give rise to currents and even conceivably electrical discharges in the polar regions.

So far, we have discussed electric fields from the theoretical standpoint, but the same important eqn (2.23) has a vital bearing on the making of measurements. A spacecraft is, after all, a moving frame of reference and the field it sees depends strongly on its velocity through the ambient magnetic field. This effect is illustrated in Fig. 2.12, where the current (actually the logarithm of the first derivative of the current with respect to voltage) which was collected by a probe, extended on an arm 1 m long from the satellite Ariel 1, is plotted. This curve shows a cyclic modulation (on the part which is a function of probe voltage) which occurs at the spin frequency of the satellite and arises from the

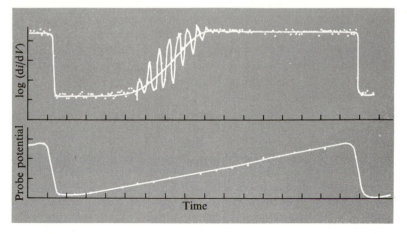

Fig. 2.12. Induction of an e.m.f. in the extended arm of Ariel I (see Fig. 1.13) and its effect on the probe current (Mullard Space Science Laboratory).

cyclic algebraic addition to the probe voltage of the e.m.f. induced in the extended arm. Before analysing the probe curve, a correction for this induced e.m.f. had to be calculated and made. It is easy to see that any effort to measure ambient electric fields by (say) measuring space potential at a pair of probes extended on arms from a satellite, would encounter the need for very accurate allowance for the induced e.m.f. due to the spacecraft motion. For example, in the auroral zone the field is about 5×10^{-5} T and spacecraft velocity may be taken as typically about 10^4 m s^{-1} for a satellite or 10^3 m s^{-1} for a sounding rocket. If an error in the electric field normal to the magnetic field of 1 mV m^{-1} is acceptable, the error in knowing the angle between the vehicle motion and the magnetic-field vector must be not more than 0·1° for a satellite or 1° for a rocket moving roughly in the magnetic field direction. If the velocity is primarily across the magnetic field, the angle becomes less critical but accuracy in knowledge of the field becomes more important, being 1000 γ for a rocket and 100 γ for a satellite.

Obviously, near the Earth, where the field lines rotate with the Earth, the electric field in the frame of reference of the rotation is what is required to enable us to understand the field's influence on the ionospheric plasma. Further out in the magnetosphere, however, the situation is more complicated since the field distribution is, to a first order, fixed with respect to the Sun and rocked diurnally by the declination of the magnetic polar axis.

In spite of the difficulties some success has been achieved in the measurement of electric fields from space vehicles by probe techniques. In these practical studies the care and attention needed to eliminate errors due to contact potential differences between two probes has posed a comparable problem to that of determining induced e.m.f.s. Probes which have a different history of manufacture or handling may readily introduce contact potential differences of many millivolts.

Because of the problems discussed above, considerable emphasis has been placed on the study of ionospheric and magnetospheric electric fields by tracking the motion of clouds of ionized vapour. The vapour commonly used is barium, which can be released from a *thermite* bomb or ejected from a thermite roman candle. Because of its low ionization potential, barium is readily ionized both by the solar radiation and by chemical reactions occurring in the burning thermite. Any vapour from the barium which is not ionized, or sometimes from strontium which may be added and is less easily ionized, forms a cloud glowing with its resonance radiation in the sunlight, while the ionized component glows with its distinctive red. The neutral component diffuses uninhibited by the magnetic field, while the ionized component is inhibited in diffusion across the field and spreads out in a cigar shape. The displacement of the ionized cloud from the neutral may be interpreted in terms of the drift motion caused by electric fields normal to the magnetic field and direction of drift. Fig. 2.13 depicts the behaviour of a cloud of such vapour ejected from a high-altitude sounding rocket. The drift of the elongated ionized component away from the un-ionized cloud in the crossed electric and magnetic fields is clearly evident.

The Sun–Earth relationship

While in Chapter 1 we were concerned mostly with thinking about the production and study of the ionized envelope of the Earth at altitudes at which the particles made frequent collisions between themselves and with the neutral gases of the upper atmosphere, in this chapter we have concentrated on the situation further away from the Earth. In the ionosphere the particles were mostly thermal, but in the magnetosphere, as we have seen, the magnetic field of the Earth retains in its neighbourhood energetic particles which would otherwise escape to interplanetary space or, if it were not for the magnetic mirror dive into the atmosphere and loose their energy there. We have seen too that the complex of magnetic field and nighly-conducting plasma is distorted and subjected to varying compression by the solar wind and its changes and that the lines of force stretching out in the Earth's wake reconnect and move, giving up their magnetic energy to trapped particles. These particles are responsible for the aurorae and indeed for many other phenomena which we have not discussed.

The energy for the production of the ionosphere comes from the Sun's electromagnetic radiation, while that for the shaping and distortion of the magnetosphere comes from its corpuscular radiation. In the next chapter we shall look briefly at the nature of the Sun and the way in which it gives rise to these two kinds of radiation. We shall emphasize especially the stormy

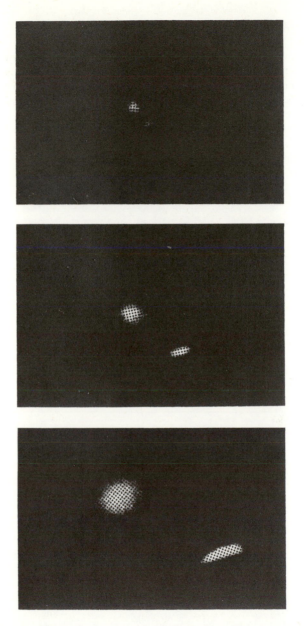

FIG. 2.13. Successive photographs of a sunlit cloud of barium and strontium vapour ejected from a rocket while the sky is dark. The glow grows by diffusion and drifts in the wind, but motion of the ionized component is inhibited except along the magnetic field lines (Föppl *et al.*).

behaviour of this star which, to our unaided sight, seems such a steady lumin-
ary. This emphasis will be made both because of the extensive range of solar–
terrestrial interactions manifested by solar disturbances and also because of
the considerable contribution to their elucidation made by the use of space
techniques.

It will be especially interesting to note some similarities between physical
processes occurring in the plasma of our own planetary envelope and those
occurring in the atmospheres of the Sun.

3. The atmosphere of the Sun

Why study the sun?

THERE are very good reasons, both geophysical and astrophysical, for studying the nature and behaviour of the Sun and important practical requirements why some of these studies need to be made from space vehicles. The Sun, as we know from daily experience, exerts an all-important influence on our (and its) planet Earth, an influence which becomes less regular and even capricious as we ascend into the upper atmosphere and out through the magnetosphere to the interplanetary medium. We have seen something of the effects of energetic corpuscular and electromagnetic radiation in the preceding chapters on ionospheres and magnetospheres. Geophysical studies therefore are inexorably bound up with solar studies, and at spacecraft altitudes show secular effects which reflect the stormy character of the Sun's own atmosphere.

But the Sun is also our nearest star, by a factor of some 300 000. Indeed, it is the only star whose disc can be seen and whose changing spectral features can be directly spatially resolved, and so there are also strong astrophysical incentives to study it. It is, in fact, a rather normal, middle-of-the-road star, obtaining its energy from the nuclear synthesis of helium—a process in which, at the high temperatures ($\sim 1.5 \times 10^7$ K) in the interior, some protons have sufficient energy to overcome their mutual repulsion and, in a succession of nuclear interactions, to build up helium atoms with the release of 6.3×10^{14} J kg^{-1} of material converted. The rest energy of a kilogram of matter is, according to Einstein, $mc^2 = 1 \times (3 \times 10^8)^2$ J, so that the efficiency of conversion of matter to energy is $6.3 \times 10^{14}/9 \times 10^{16} = 0.7$ per cent.

The total power radiated from the Sun is 3.9×10^{26} W which indicates that hydrogen is being consumed at a rate of 6×10^{11} kg s^{-1}.

This large flow of energy forces its way out by convection and radiation and absorption and re-radiation of the various photons very many times in the journey. As in most other stars, it is the enormous pressure of this radiation which keeps the Sun *inflated* against the huge contracting force of its own gravity, so that its mean density is only that of water. In this context it is interesting to note analogously that the major mechanical force in the close neighbourhood of a thermo-nuclear explosion is the pressure of X-radiation, as can easily be shown. A photon of frequence v has an energy hv and momentum hv/c. The energy flux from a dense (black-body) plasma is given by Stefan's law as σT^4 (where $\sigma = 5.67 \times 10^{-8}$ W m^{-2} K^{-4}), and so the photon flux is $\sigma T^4/hv$ photons per square metre and the momentum flux (pressure of radiation) is given by $(\sigma T^4/hv)(hv/c) = \sigma T^4/c$. At a distance R from a fireball of radius r the pressure will be $p \sim \sigma T^4 r^2 / c R^2$. Inserting $T = 10^8$ K, $r = 0.1$ m, and $R = 100$ m, we obtain $P \sim 1.9 \times 10^{10}$ N m^{-2} (which is about 1200 ton in^{-2}).

The visible 'surface' of the Sun, known as the *photosphere*, is no more solid than the rest of the gaseous ball, but it seems to be a surface because it is that thin shell of the Sun's bulk from which photons of visible light can escape without a high probability of absorption and re-emission. To put it another way, visible-light photons from the photosphere reach us for the most part without encountering atoms on the way, which is why it is the surface which we see. In other wavelengths, notably ultraviolet and X-rays, the surface would appear to be at a greater solar altitude. In certain individual spectral lines in the visible, whose photons have a large probability of interaction with the solar atmosphere, the surface would appear higher than the white-light surface. With a good telescope, in good seeing conditions, the photosphere is found to have a mottled appearance. These *granulations* are convection cells, with hot material rising in their bright centres and falling at their somewhat duller boundaries. The power to drive this convection comes from a celestial heat-engine cycle in which elements of the gas, consisting largely of hot protons, electrons, and hydrogen atoms, which happen to be rising and to be near the outer part of the Sun, find themselves in an unstable condition in which their bouyancy increases as they rise, which results in a turbulent motion. This situation is somewhat analogous to the turbulent motion which gives rise to weather in our own troposphere, when rising elements of air find themselves in regions of greater density than that attained by the elements by adiabatic expansion of the elements of air themselves. The network of photospheric granulations is contained in a larger-scale convection network of *supergranulations*, whose boundaries enclose about 10^3 granulations and are delineated by a small increase in the solar magnetic field and by hedge-like bushes of *spicules*—short-lived jets of cooler, denser gas penetrating into the upper atmosphere of the Sun.

Looking into the solar atmosphere (Fig. 3.1)

We have already seen that the depth we see into the Sun's atmosphere depends on the wavelength of the light employed. We use this fact, by looking at the Sun in various wavelengths, to study regions at various heights above the photosphere. Because atoms and ions have a large cross-section for absorbing radiation at the frequency of their resonance lines (lines which are absorbed by exciting the atom or ion from its ground state to a higher level), these lines characterize greater altitudes. However, there is another reason why the more energetic photons are seen to come from greater altitude. The outer atmosphere of the Sun—known as the *corona*—has a temperature of around 1.5×10^6 K. At these temperatures many atoms are multiply-ionized, so that we find far-ultraviolet spectra of abundant atoms like iron with many

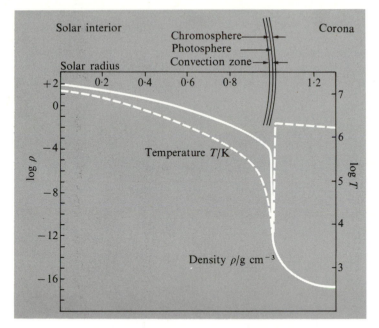

FIG. 3.1. Diagram showing the variation of the logarithms of temperature (log T) and density (log ρ) with radial distance from the centre of the Sun in units of solar radii. The position of the Sun's visible surface—the photosphere—with the convection zone below it and the chromosphere and corona above are also shown.

or all but one of their electrons missing. This comes about because, at these temperatures, the electrons in the coronal plasma have the energies of hundreds of electron volts necessary to produce these high degrees of ionization.

The need for spacecraft

Since the flux and wavelength of the various solar photons that reach the neighbourhood of the Earth convey information about the physical conditions obtaining at the altitudes in the solar atmosphere from which the photons have come, ultraviolet and X-ray studies provide us with a diagnostic tool for discovering the conditions in the extensive plasma surrounding the Sun. To do this, space vehicles are necessary in order to overcome the first of the three classical limitations to ground-based astronomy—absorption by the Earth's atmosphere (see Fig. 3.2). But the second and third limitations—irregular refraction of light, which gives rise to poor astronomical *seeing*, and atmospheric scattering of sunlight (or light from other sources)—are usefully conquered also. Increasingly, the improved stability of space-borne telescopes is creating a situation in which we can expect to improve on the

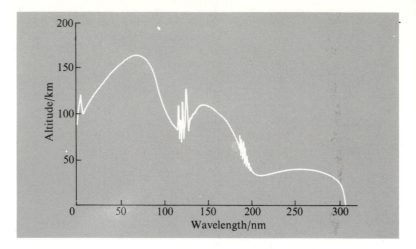

FIG. 3.2. Transparency of the atmosphere. Profile of unity optical depth as a function of wavelength looking vertically into the earth's atmosphere, that is, the altitude at which zenith radiation of a given wavelength has been attenuated by a factor e.

spatial resolution available for the study of structures in the solar atmosphere, which from the ground is normally around 10^3 km. Already, white-light *coronographs*, by forming an artificial eclipse, have made it possible to study the faint light of the solar corona at several solar radii, without dependence on that rare event, a total eclipse of the Sun. At ground level the performance of a coronograph is greatly limited by light scattered in the Earth's atmosphere, which swamps the faint coronal light at a short distance from the solar *limb* (the edge of the disc). We can realize why this happens as soon as we remember that our name for the scattered light is the blue sky.

No part of astronomy stands alone. Each wavelength range makes its contribution to the overall understanding of the system or class of object under consideration. This is as true for the Sun as for any other star or cosmic object. Although it is not possible in the scope of this chapter to describe the vital contribution to our understanding of the Sun that comes from terrestial studies, special mention must be made of the results from radio astronomy. The energetic particles that result from the occurrence of a solar flare give rise not only to outbursts of X-rays but, by exciting oscillations in the coronal plasma and by generating synchrotron radiation, they also produce ratio outbursts. The study of these various kinds of burst has played an important role in solar physics. In the visible too a great deal of work has been, and continues to be, done; and pictures from terrestrial observatories taken in various spectral lines provide an essential daily monitor of conditions at various levels above the photosphere.

While space astronomy of the Sun has played a significant role in solar physics, especially in establishing the temperature–altitude profile of the atmosphere of the quiet Sun, it is in elucidating the behaviour of the active Sun that the contribution of observations above the atmosphere has been greatest.

Solar activity

There is a very close connection between the activity of the Sun and its magnetic field. Roughly every eleven years—the so-called *sunspot cycle*— the magnetic field of the Sun reverses in sign. In the intervening years, the fact that the equatorial rotation rate of the highly-conducting fluid of the Sun slightly exceeds the polar rate results in the creation of a generally circumferential field, which here and there, in middle solar latitudes, penetrates the photospheric surface and gives rise to a pair or more of sunspots of opposite magnetic polarity. Typically, these spots have a diameter of about 20 000 km and a magnetic field of several tenths of a tesla. These fields inhibit the transverse migration of electrons upon which the thermal conductivity of the medium chiefly depends, and so the cores of the sunspots in the photosphere appear dark, being cooler than their surroundings.

The regions above the photosphere around sunspots, or where sunspots are likely to appear, are hotter, denser, and more magnetic than the regions of the quiet Sun. This activity is most evident in the *chromosphere*, an envelope of atmosphere some 10^4 km thick, through which the temperature rises from a value a little less than that of the underlying photosphere toward the million or so degrees of the solar corona. The higher temperature of these active regions is probably connected with the effect of the enhanced magnetic field on the heating action of the mechanical waves propagated upward from the photospheric granules.

Since, as we have already seen, the hotter the plasma the more it radiates at short wavelengths, it is clear that the flux of high-energy photons, which convey information from the higher, hotter parts of the solar atmosphere, will be enhanced when solar activity increases either the density or temperature (or both) of these emitting regions. On the other hand, data on the photosphere are largely conveyed by visible photons, the dark Fraunhofer absorption lines characterizing the cooler region just above the photosphere, in which they are formed by absorption. The nearer, ultraviolet radiation is diagnostic of the chromosphere and broadly of quiet conditions.

Although much beautiful work has been done on solar ultraviolet spectroscopy and spectroheliography (pictures in spectral lines), only one example of instrumentation will be described, since it will illustrate many of the basic principles. The now often more sophisticated instrumentation employed is not otherwise especially characteristic of space work, and the interpretation of the spectra is a highly specialized study which cannot be developed here.

An ultraviolet solar spectrograph

Early in the history of space science it was discovered, using V-2 rockets, that the continuous visible spectrum of the Sun covered with Fraunhofer absorption lines gives place in the ultraviolet to a diminishing continuum with very prominent emission lines. Since those first flights, many studies have been made of the Sun's disc in ultraviolet light.

The first ultraviolet spectra to be obtained from beyond the *limb* were obtained in 1965 by a simple concave-grating spectrograph flown on a stabilized Skylark rocket from Woomera. Fig. 3.3 is a diagram of the instrument. The nose-cone of the rocket was ejected, and the instrument bay separated from the solid-fuel booster. Gas jets of the primary stabilization system, controlled by signals from photocells viewing the Sun itself and by a sensing of the Earth's magnetic field, pointed the bay towards the Sun with a precision of a minute or so of arc. Secondary stabilization to within a few seconds of arc of the solar vector was produced by a pair of concave mirrors fixed rigidly on a servo-driven mount capable of small movements about two axes normal to the solar vector.

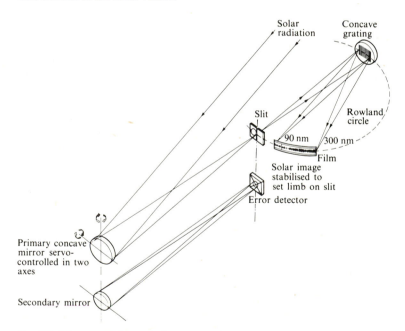

FIG. 3.3. Diagram of the optical system used to photograph spectra from the solar chromosphere beyond the limb from a Skylark rocket. Note the secondary mirror and error detector to provide signals to servo-control the pointing of the primary mirror on the same mount (Burton, Ridgeley, and Wilson 1967).

The two mirrors each formed an image of the Sun, the secondary on a quadrant split-field error-detector feeding the servo. The output of the servo was so arranged as to maintain the image of the Sun due to the primary on one jaw of the spectrograph slit such that the slit was set just above the image of the solar limb. The light from the Sun's chromosphere, after passing through the slit, fell on a concave grating which focused a spectrum on to a piece of curved film mounted in a multi-exposure camera. Part of a spectrum obtained is shown in Fig. 3.4, together with the identifications of the elements from which

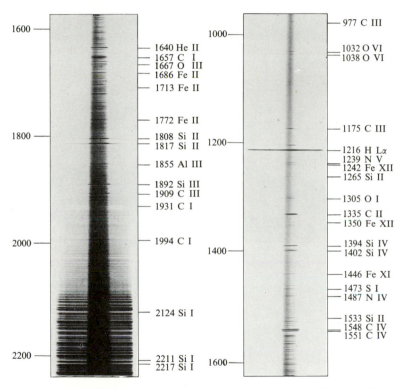

FIG. 3.4. Spectrum from just above the solar limb showing emission lines of multiply-ionized elements in the chromosphere. Wavelengths in ångstrom units, that is, 10^{-1} nm (Burton *et al.* 1967).

the line arise. The most pronounced line is the resonance radiation of the hydrogen atom at 121·6 nm. Most of the other lines are from elements in multiple stages of ionization.

Measurements of this type, in which the intensities of the lines are carefully measured by a microphotometer and in which the slit position above the limb is varied, provide a more precise method of studying the temperature structure of the chromosphere than can be obtained by observations of the disc. Nevertheless, when it is realized that the chromosphere is far from horizontally uniform being threaded through by spicules, it is evident that future technology must produce a system able to resolve the temperature structure of individual spicules a second of arc in diameter.

Obviously the advantages of photography are not readily available, except with recoverable rockets or satellites. To date, much work has used photoelectric detection, which provides a signal suitable for telemetering and has the advantage of greater linearity of response and dynamic range. Increasingly,

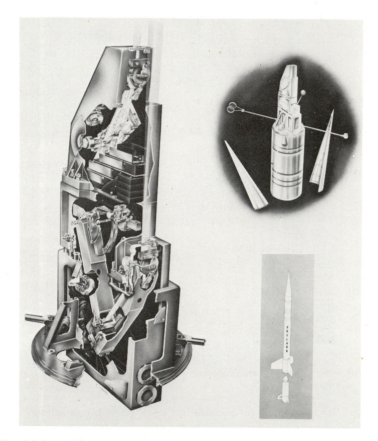

FIG. 3.5. Pair of fixed and scanning extreme ultraviolet grating spectrometers for a study of the atmospheric absorption of solar radiation (Mullard Space Science Laboratory).

elegant television devices are employed, but it is very difficult to match the huge information-capacity of photographic film. Fig. 3.5 illustrates a rocket payload consisting of a scanning photo-electric spectrometer together with one operating in four fixed wavelengths.

The Sun as an X-ray star

A star like the Sun is remarkable for two main reasons. First, the relatively cool photosphere (as stellar surface temperatures go) happens to lie in the range of temperatures in which the hydrogen is only partly ionized. This gives rise to the turbulence, evidenced by the granulations, and this in turn heats the corona to nearly 2×10^6 K by the dissipation of mechanical energy. We have therefore the strange phenomenon of a relatively cool body in space enveloped in an immensely hot atmosphere. (We can note in passing that the Earth's upper atmosphere is hotter than its surface but this is less remarkable as in the Earth's case the energy comes from without.) Secondly, the differential rotation of the Sun combined with its magnetic field, which no doubt results from a self-excited dynamo motion of the solar fluid driven by the angular momentum of the rotation, gives rise to the solar cycle of activity. This shows itself most clearly in the occurrence of sunspots, together with a whole range of phenomena associated with active regions.

The net result of these two aspects of the Sun's behaviour is, first, that it radiates significantly in the soft X-ray part of the spectrum (say wavelengths beyond 10 nm or photon energies greater than 123 eV) and, secondly, the X-ray flux hardens and increases as activity results in higher-temperature and denser regions in the corona. This amounts to saying that the Sun is a variable X-ray star; it is fortunate for us that the variability is not reflected in the energy flux in the visible.

Fig. 3.6 shows a photograph of the Sun in white light taken during an eclipse. Even before the first discovery of X-rays from the Sun, by instruments flown on captured V-2 rockets, observations of this kind suggested that very high coronal temperatures would be necessary to support such a distended atmosphere, and the structured nature of the plasma hinted at the influence of magnetic fields. It turns out that the soft X-ray spectrum of the quiet Sun has a maximum around 3 nm, and so, if for the moment we think of the corona as radiating like a black body, Wien's displacement law ($T \sim 0.003/\lambda_{max}$) would suggest a temperature of 10^6 K. A 'black-body radiator' is one in which the frequency spectrum of its energy is randomized by many absorptions and re-emissions, before finally emerging with a Planck spectral distribution (first calculated by Planck on the basis of quantum statistics alone). In fact, the corona is optically thin to the X-radiation; photons are not rattled around in it by much absorption and re-radiation before escaping, so it does not radiate as a black body and, as we have seen, its temperature is rather more than 10^6 K.

FIG. 3.6. Coronal photograph from the 1966 expedition of the High Altitude Observatory. High Altitude Observatory is a Division of the National Center for Atmospheric Research, Boulder, Colorado. The National Center for Atmospheric Research is sponsored by the National Science Foundation.

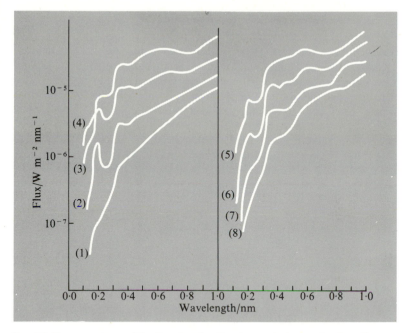

Fig. 3.7. X-ray spectrum of the Sun showing the change during a solar flare. Spectra were obtained by the instrument of Fig. 3.9 and cover a period of 25 min (Culhane, Sanford, Shaw, Phillips, Willmore, Bowen, Pounds, and Smith 1969).

Fig. 3.7 shows the X-ray spectrum of the Sun during a 26 min period on 26 October 1967. In this interval the phenomenon of a *solar flare* occurred; a region of the disc covering rather less than 10^9 km^2, that is, less than 0·025 per cent of the visible hemisphere, suddenly brightened, especially in radiation associated with excitation of hydrogen atoms which characterizes the bottom of the chromosphere (flares are normally monitored in the H-α line of hydrogen at 656·3 nm). In spite of the tiny fraction of the disc over which the visible event occurred, the X-ray flux over most of the observed range increased by between one and two orders of magnitude. This behaviour is quite typical of solar flares, which are fairly frequent events especially at the maximum of the solar cycle. Although some of the radiation may come from non-thermal processes—*bremsstrahlung* from energetic electrons—much of it is due to the sudden heating of coronal material to $3–4 \times 10^7$ K. Considering the small region of the Sun involved, the energy emitted during this period is enormous and can only be accounted for in terms of the magnetic energy associated with the fields of sunspots. We will return to this after describing the instrumentation.

A proportional-counter solar X-ray spectrometer

The spectra of Fig. 3.7 show sufficient structure to suggest the presence of unresolved spectral lines. There is obviously a feature due to a line or group of lines at about 0·2 nm. A number of important studies of the solar X-radiation have been made using Bragg crystal spectrometers to resolve the line structure, and the fact that the spectrum is primarily of this kind and not a Planck black-body continuum is another aspect of the optical thinness of the corona. It is easy to see that this has to be the case by calculating the energy which would be emitted as a black body. Stefan's constant is $5·74 \times 10^{-8}$ W deg^{-4} m^{-2}, so that the radiation from a corona 10^6 km in radius at a temperature of $1·5 \times 10^6$ K would be $4\pi(10^9)^2 \times 5·74 \times 10^{-8} \times (1·5 \times 10^6)^4$ W $= 3·7 \times 10^{36}$ W, that is, 10^{10} times the total solar power! Fig. 3.8 is a spectrum from

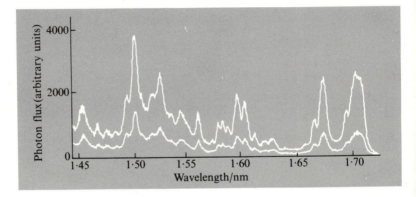

FIG. 3.8. X-ray spectrum from solar active region, resolving individual lines, mostly of iron; obtained with a Bragg crystal spectrometer (Mullard Space Science Laboratory).

an active region obtained by a payload consisting of a pair of Bragg crystal spectrometers carried by a Skylark rocket. However, until recently, most work on X-rays from solar flares, which because of their short life and unpredictability require the use of satellites rather than sounding rockets for their study, has made use of the proportional gas-filled photon counter. In its simplest form such a detector consists of a chamber filled with gas at a pressure of about 10^5 N m^{-2} and fitted with a thin window to admit X-rays. The centre of the chamber is spanned by a fine wire about 40 μm in diameter at a positive potential of a kilovolt or so.

Suppose a photon of energy 2 keV (wavelength $\sim$0·6 nm) passes through the thin beryllium window of such a device and interacts with the gas. It will eject a photo-electron (δ-ray) which will ionize the gas, resulting in perhaps 60 ion–electron pairs. The electrons drift towards the anode, where they encounter the very high field in its neighbourhood and produce an avalanche of ionization resulting in a pulse of charge arriving. Although the *gas amplification* may be as much as a factor of 10^5, the charge in the pulse remains proportional to the energy deposited by the original photon. By employing suitable electronic circuits, the number of pulses in a given series of amplitude ranges may then be counted and the spectrum of the photon flux found.

Fig. 3.9 depicts the instrument flown on the Orbiting Solar Observatory OSO-4, with which the spectra of Fig. 3.7 were obtained. The extensive electronic pack would nowadays occupy only a few square centimetres of circuit board—such has been the improvement in miniaturization made possible by integrated circuits. The instrument carried an array of counters with various window materials and sizes to cover the spectrum from 0·1 nm to 5·6 nm.

The need for a number of different counters with different window materials and filling gases arises from the requirement to optimize the window transmission and the absorption of the photon in the gas over a number of wavelength ranges and, in the case of the harder ranges, over a considerable dynamic range of intensities, because of the huge changes in flux during a solar flare.

The energy of a solar flare

A typical flare of the kind observed by OSO-4 and whose occurrence is portrayed in Fig. 3.7 might occupy an altitude range of 10^4 km in the upper chromosphere and lower corona. If the area is taken as 10^9 km^2 the volume involved is 10^{13} km^3. It is instructive to calculate how much thermal energy there would be in such a volume. The thermal kinetic energy density is $\frac{3}{2}nkT$, where n = number of particles per cubic metre, k = Boltzmann's constant = $1·38 \times 10^{-23}$ J K^{-1}, and T = absolute temperature. Although n and T vary considerably throughout the volume, their product remains of order 10^{21} m^3 K, and so the energy in 10^{13} km$^3 \sim 10^{21} . 10^{-23} . 10^{13} . 10^9 \sim 10^{20}$ J, but the power emitted by the flare in X-rays alone is 100 times this value. It is therefore evident that the energy source of the flare is not thermal. Thermo-nuclear energy would be adequate, but can be discounted on the grounds that only very rare, very large flares give an even-detectable flux of neutrons at the Earth. The remaining plausible source of energy is that associated with the magnetic field of the sunspot system.

The energy density in a magnetic field is $B^2/2\mu_0$ (which for a mean field of say 0·1 T is 4×10^3 J m^{-3} or a total energy in the 10^{13} km^3 of 4×10^{25} J). This is more than enough to give the observed X-ray flux at less than 1 per cent efficiency.

There are several mechanisms which have been suggested for the way in which the magnetic energy is released. The two main types have marked similarities to the mechanism discussed for the release of magnetic energy from the magnetospheric tail in Chapter 3, since each model conceives the energy as

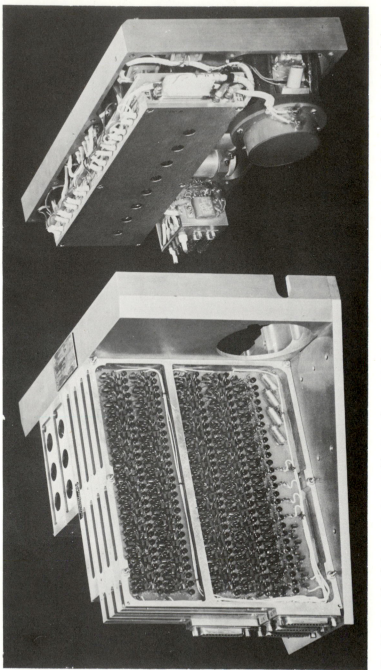

FIG. 3.9. Ensemble of photon counters sensitive to various spectral ranges together with an electronic pack of the pre-integrated circuit days (Culhane *et al.* 1969).

resulting from the annihilation of the field as regions of opposing polarity approach, compressing the entrapped plasma. The models differ in the way in which they envisage the opposing fields as arising, and indeed each may be valid in different circumstances.

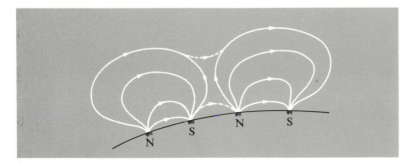

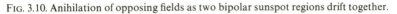

FIG. 3.10. Anihilation of opposing fields as two bipolar sunspot regions drift together.

In the first model (see Fig. 3.10) the opposing fields are seen as caused by two pairs of bipolar sunspots which drift together, compressing and heating the plasma. A triggering action results from the high magnetic pressure gradient on each side of the neutral sheet; this may be thought of as arising from the property of repulsion attributed to neighbouring lines of force of the same sign and of attraction and linking-up attributed to lines of opposite sign. A glance back at Fig. 2.2 will illustrate like field lines repelling (over the poles) and a consideration of the behaviour of two oppositely oriented magnets brought together sideways will recall the tendency of the two suddenly to jump together as the force increases rapidly with falling separation. Even when this is taken into account, the process would be far slower than the observed rate of rise of emission in a flare were it not for the generation of magneto-hydro-dynamic waves which provide a mechanism for the rapid conversion of magnetic energy to heat. This model can account for flare spectra of the thermal type, where the emission is from hot plasma.

In some flares the spectrum is non-thermal and more characteristic of X-rays generated by streams of energetic particles. The second model (Fig. 3.11) seeks to account for these streams by supposing that the solar wind draws out the field from a single bipolar sunspot pair, very much as the magnetospheric tail is drawn out. Just as, in that case, reconnection of the field lines results in fast particles causing auroral sub-storms, so in this model fast electrons are injected into the denser chromosphere, where they give rise to an impulsive burst of non-thermal X-rays.

FIG. 3.11. Diagram showing how field lines from a sunspot pair drawn out by the solar wind may reconnect releasing energy and generating fast particles which are trapped on the lines and impact on the denser lower solar atmosphere. (After Strauss and Papagiannis 1971.)

An X-ray 'spectroheliograph'

Because the occurrence of a solar flare out-shines the rest of the Sun by a large factor in the harder part of the X-ray spectrum, these events may be studied without telescopes to select their flux from the background emission. The study of active regions, however, requires the ability to dissect the image of the Sun. Such an instrument was flown on OSO-5. A telescope in the form of a single parabolic X-ray mirror, of the kind described in the next chapter, was mounted in the pointed arm of the satellite (Fig. 3.12). This arm scanned the Sun in a raster, so that X-rays from a succession of elements of the solar disc were directed into a proportional counter at the focus of the mirror. Contour maps of the Sun were constructed from these data, and published daily for nearly four years, providing a means for studying individual areas of activity as they formed, decayed, or crossed the solar disc.

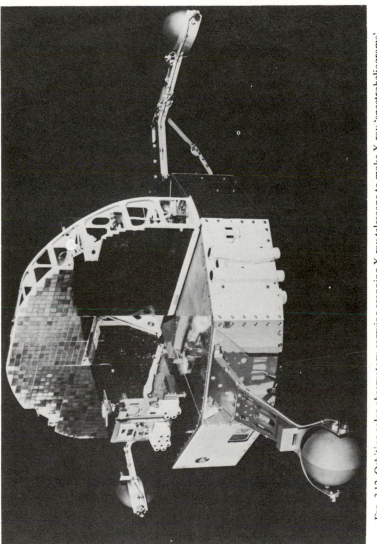

FIG. 3.12. Orbiting solar observatory carrying scanning X-ray telescope to make X-ray 'spectroheliograms' (National Aeronautics and Space Administration).

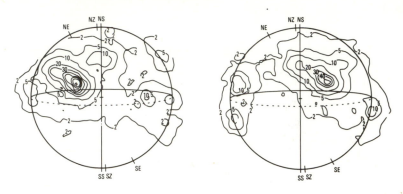

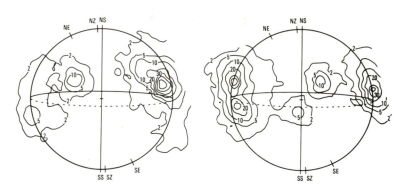

FIG. 3.13. X-ray spectroheliograms obtained by the instrument on OSO-5 on the 1, 4, 7, and 9 April 1969. The wavelength range for these maps was $0.91–1.05$ nm and the contours run from 2×10^{-9} Wm^{-2} to 40×10^{-9} W m^{-2} at the Earth's orbit (Mullard Space Science Laboratory).

A typical series of maps spanning a period of nine days showing a large active region crossing the disc is given in Fig. 3.13.

Skylab

At the time of writing, the first results are just coming in from a venture which represents an important step forward in the technology of space astronomy. The huge developments and capabilities resulting from a con-comitant expenditure on the Apollo lunar adventure has given scientists facilities for solar studies which would certainly never have been made available for that reason alone. The Skylab project has sought to provide, largely from components used in the Apollo lunar work, a manned solar observatory in space.

While the equipment was not planned to make the maximum use of the potentialities of the astronaut's presence, and indeed in several cases could be operated entirely automatically, for one reason or another the value of man's presence in an operation of this kind, in the view of some former sceptics, has been amply justified. This was demonstrated especially by a failure of a solar battery to deploy and the need to prepare, transport to the spacecraft, and erect a special sunshade to maintain correct temperature in spite of the malfunction.

A list of the experiments carried serves both to show the current state-of-the-art and to illustrate the kind of observation most important in the present phase of solar studies from space. Apart from a visible-light telescope, operating in the H-α light of hydrogen atoms to monitor the details of the solar weather (see Fig. 3.14) and especially to reveal the occurrence and location

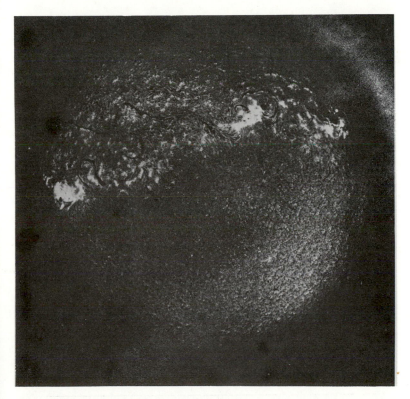

Fig. 3.14. Photograph of the Sun in H–α (656·3 nm) obtained from a terrestrial observatory by Dr. Weiner Neupert, Solar Physics Branch, Goddard Space Flight Centre, NASA.

of solar flares and active regions, all the astronomical instruments but one were concerned with ultraviolet or X-ray observations. There were two X-ray telescopes capable of producing true images of the Sun, one of them with a transmission diffraction grating giving images in the range 0·2–6·0 nm, with a resolution better than 0·1 nm, and the other arranged to give high time-resolution photographs in the 0·6–3·3 nm range. These radiations characterize, as we have seen, the corona of the quiet Sun and the chromosphere and lower corona during disturbances. Fig. 3.15 is a photograph obtained with one of these instruments, and may be compared with Fig. 3.13 to illustrate the improvement in telescope technique (see Chapter 4) and the value of recoverable photographic film.

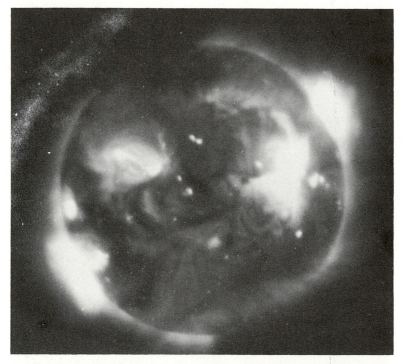

Fig. 3.15. X-ray photograph of the solar corona obtained from Skylab on 28 May 1973 by the Solar Physics Group at American Science and Engineering (Vaiana, Davis, Giacconi, Krieger, Silk, Timothy, and Zombeck 1973).

There was a whole-disc spectrometer giving a series of images of the disc in the principal spectral lines which arise in the corona in the range 15–62·5 nm, and there was a photographically-recording double-grating spectrograph to obtain highly dispersed spectra of selected areas of the disc in the range 97–394 nm which characterizes the chromosphere. The exception to ultra-violet and X-ray instrumentation was a coronagraph operating in white light, in which an artificial eclipse occulted the visible disc and a region a quarter of a radius beyond it and so allowed a series of photographs to be taken, showing up the coronal streamers and their movement. The large part played by photographic recording implies a major role for the astronauts in attending to the cameras.

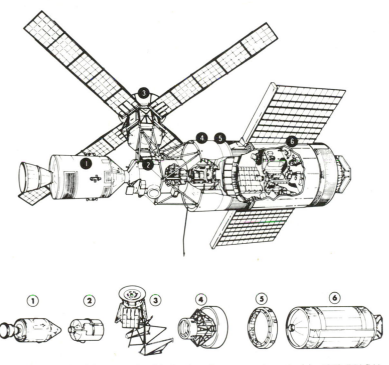

FIG. 3.16. The Skylab manned orbital solar observatory launched in 1973 (NASA). 1. Command and service module; 2. Multiple docking adapter; 3. Apollo telescope mount; 4. airlock module; 5. instrument unit; 6. orbital workshop.

Fig. 3.16 is a diagram of this sophisticated space observatory, showing its major components and their function. Several successive visits to the observatory, lasting many weeks, have taken place and in this way have set the style for much of the space science to be conducted in the next decade by the space station and shuttle system. It should be mentioned perhaps, that experiments in fields other than solar physics are also being carried out and that some useful work is possible during the unmanned period.

4. Cosmic X-ray astronomy

Beyond the X-ray curtain

FOR centuries the only information reaching mankind from the domain of the stars came as the flux of visible photons collected by the 7 mm diameter pupil of the human eye. Then for three more centuries the flux-gathering capacity slowly grew, as bigger telescopes were built, though the spectral range increased but slightly by extension into the near ultraviolet and infrared, using photographic emulsions and bolometers as detectors. After the Second World War radio astronomy greatly increased the spectral range, and then, with the advent of space vehicles, ultraviolet astronomy was developed. The significance of cosmic ultraviolet studies lies principally in the fact that the resonance radiation of most elements in both their neutral and ionized states occurs in the ultraviolet. Since resonance radiation is commonly the most intense emitted its diagnostic value is great, but what of the X-ray part of the spectrum? What value might it have?

We saw in the last chapter that X-ray emission characterizes plasmas at millions of degrees. Therefore in the early 1960s, when the first tentative steps were being taken, the likely significance of cosmic X-ray astronomy turned largely on the possibility of the occurrence of such plasma in observable quantities in the cosmos. Before following this quest further we must take note that there is a gap in the observable spectrum between about 10 nm and 90 nm, which occurs because of absorption in the interstellar medium. This opaque region does not interfere with solar astronomy, since at the short distance involved the interplanetary gas is not sufficiently dense to obscure the Sun at these wavelengths, but at the huge distances of the stars only the very nearest objects could be seen in this light.

We might expect to learn what could be observed in X-rays by considering lessons learnt from the Sun.

1. X-rays are emitted by very hot plasma (cf. Wien's law) or non-thermal emission processes due to energetic electrons.
2. Since Stefan's law, which applied to power emission from a dense plasma behaving as a black body (see Chapter 3, p. 53), indicates the power flux to be proportional to T^4, the emitting plasma for any reasonable power input must either be small or optically thin. An optically-thin plasma is one in which photons escape—on the whole—as they are generated, without absorption and re-emission. The flux from such a hot plasma falls far below that given by Stefan's law. At 6×10^6 K a plasma the size of the Sun and dense enough to radiate as a black body would require 10^{12} times the solar energy output to maintain its temperature.

3. Since the energy flux from even a small blob of very hot plasma is enormous, rapid heating and cooling can take place so that fast varying phenomena may be expected (cf. solar flares).

To some extent these ideas, though simple, are being wise after the event. In 1962, before any cosmic X-ray source had been discovered, all that was known for certain was that, were the Sun to be placed (in imagination!) at the distance of a rather near star, our techniques would be unable to detect it. It was therefore with hope rather than certainty that the first explorations were made.

Early discoveries

In the early 1960s two groups in the U.S.A., one under Giacconi at the American Science and Engineering Corporation and the other under Friedman at the American Naval Research Laboratory, and the author's group in the U.K. were planning to look for cosmic X-ray objects, indeed the Naval Research Laboratory had already had a rocket flight, which, although it appeared ambiguous at the time, can be seen now to have detected a cosmic flux; but, such was the atmosphere of scepticism, that the payload which did in fact discover the first cosmic source was intended, at least nominally, to look for X-ray fluorescence from the Moon. This rocket (Fig. 4.1), flown by Giacconi and his colleagues, obtained the results shown in Fig. 4.2.

Clearly there is a strong signal which does not quite line up with the lunar direction as the field of view was carried past the Moon by the roll of the rocket. However, the direction of the peak was closely aligned to the direction of the Earth's magnetic field, and until a further flight was made there remained some uncertainty as to whether the count detected might arise from penetration of the counter by magnetically trapped particles or by X-rays produced by them. We shall see that dealing with 'noise' background such as this is one of the major technological problems of cosmic X-ray astronomy.

In the event, the object responsible for the large signal turned out to be the source Scorpius X-1 (Sco X-1)—the first sources were named after the constellations containing them and the order of their discovery in that constellation. Sco X-1 was an abundant justification of the hopes of would be X-ray astronomers. It turned out to be radiating about 10^{30} W in the X-ray part of the spectrum alone, which is three orders of magnitude more than the total output of the Sun over the entire spectrum.

Soon after, the Crab Nebula (the remnants of a supernova explosion in the year 1054, which had for some time been suggested as a possible X-ray source) was found to be radiating 10^{31} W in X-rays, though, being further away, it was fainter. It was immediately clear that X-rays were likely to provide an insight into processes very different from those generating the energy in normal stars. Indeed, it was not long after the discovery of the first pulsing radio star

FIG. 4.1. Rocket payload which discovered the first cosmic X-ray source Sco X-1 (Giacconi and Gursky 1965).

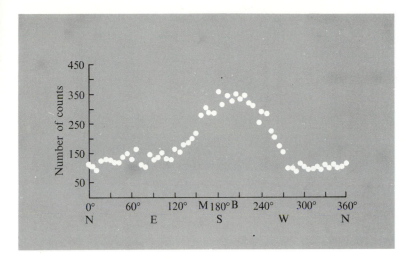

FIG. 4.2. Counts registered by one of the counters on the payload of Fig. 4.1. M is the direction of the Moon and B that of the Earth's magnetic field (Giacconi and Gursky 1965).

(*pulsar*) that the Crab was shown to contain such an object which pulsed thirty times a second, not only in radio but in the visible and in X-rays.

During the first eight years of the new astronomy, before a satellite devoted to the subject had been launched, most of the work was done with vertical sounding rockets, although in the high-energy part of the spectrum, beyond 24 keV (0·05 nm), some use was (and still is) made of high-flying balloons. Amongst the discoveries made and features which became clear in this period were the following:

1. Sco X-1 was identified with a faint blue, slightly variable star, and its distance (from which the power mentioned above was calculated) was set at around 1000 light-years from measurements of the interstellar absorption of the lines of ionized calcium.

2. Sco X-1 was found to flare in X-rays so that its strength varied by a factor of 2 or more in hours.

3. The Crab was studied by observing it during a lunar occultation. This showed the source to be distributed. The pulsed component however appears to be a point source identified with the radio pulsar.

4. An X-ray nova, Centaurus X-2 (Cen X-2), was discovered. It had not been detected in flights before 1966; in April 1967 it appeared brighter than Sco X-1, and six months later it had vanished.

Fig. 4.3. M 87 (Lick Observatory photograph).

5. Extragalactic sources were discovered, notably Virgo X-1, which was tentatively identified with the galaxy M87 (see Fig. 4.3), a strong radio source. If the identification is correct the X-ray power is two orders of magnitude greater than the radiopower.

6. The sources in our galaxy were found, as might be expected, to cluster along the Milky Way—the plane of the galaxy—especially towards the galactic centre (Fig. 4.4). An estimate of the X-ray power of the whole galaxy gave a figure approaching 10^{33} W, closely comparable with the radio power.

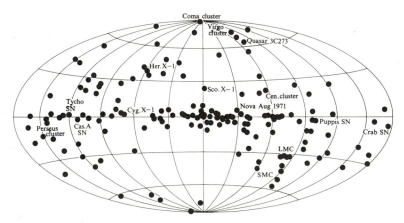

Fig. 4.4. The X-ray sky as seen by *Uhuru*, January 1973.

7. A diffuse isotropic unresolved background flux was discovered.

In addition to the features implied by the above list of discoveries, in many cases other similar sources were found, so that Sco X-1, Cen X-2, and Tau X-1 (Crab) are prototypes for classes of object.

We shall return to discuss these objects later, but first let us consider the technology with which the discoveries were made. Although balloons had a role we will concentrate on rocket techniques, as they were by far the most significant.

Large-area detectors for sky surveys

From the first it was clear that the ideal approach would ultimately involve building X-ray reflecting telescopes to be carried on highly stabilized observatory-type satellites, but in 1962 no such satellite existed and those planned (The Orbiting Astronomical Observatory Series) were to be devoted to ultraviolet astronomy, the planning having taken place before the discovery of Sco X-1. On the other hand, an observatory is little use without some indication of where to look, and the unstabilized or poorly stabilized rockets with which we entered t' decade provided a suitable platform for scanning the sky and building up a catalogue.

What was needed was the maximum area sensitive to the arrival of X-rays, means for rejecting counts produced by charged particles in the upper atmosphere, and efficient means for collimating the field of view of a detector, preferably without seriously cutting down the available observing time. It is, of course, necessary also to provide the payload with a sufficiently accurate method of finding its aspect at any instant.

The three primary needs were normally met by employing multi-anode proportional counters, so arranged as to distinguish between the arrival of an X-ray which ejects a short-range photo-electron from the filling gas and the long ionization track produced by a particle. Directivity was provided initially by a honeycomb or slat (egg-crate) collimator—a structure that, being placed in front of the counter, limited the field of view to a few degrees. More recently, a parallel-wire system known as an Oda collimator became the forerunner of a number of *binary mask* systems for coding incoming signals. A binary mask is a system of transparent and opaque regions which codes a flux in a manner that can be decoded to give its direction of origin. We shall see later that this coding can be either temporal or spatial.

The signal/noise problem

The product of collecting area by observation time is never as large as one would like. The minimum detectable signal in the case of a large-area counter is limited by possible contributions from three sources of noise. A small detector or detector element at the focus of a large mirror may suffer the same limitations, although then it may be that the noise is negligible and the statistical significance of the signal by itself is the important quantity. Obviously, if only one photon is registered during an observation period, one cannot say anything very precise about the source strength.

Let the intrinsic noise in the detector be designated by B_i counts per second. In gas counters this effect is due to residual radioactivity in the counter and is normally negligible. In solid-state and photo-electric devices, however, it has a thermal origin and must not be overlooked. The background from cosmic rays and charged particles may be designated by B_c counts per second per square metre of counter exposed to the flux. When not in the radiation belts, $B_c \sim 2000$ counts per second in an unprotected gas counter. There remains the signal from the diffuse sky background which, unless it is the object of study, represents a noise, and over the energy range below 0·3 keV (4 nm) has a value $B_d \sim 2 \times 10^5$ counts per second $sr^{-1} m^{-2}$.

Now the noise can be measured and subtracted, but the precision with which this can be done is limited by its statistical fluctuation. This fluctuation is of order $\sigma = $ (total noise counts)$^{\frac{1}{2}}$, and to make sure that a signal is real we normally require it to exceed this value by 3 (sometimes 5σ is taken and commonly the ratio (signal/σ) is used as a measure of the significance of the signal).

If T is the time for the observation, A_x is the collection area for X-rays, A_p is the area sensitive to the particle flux, s is the minimum detectable signal in counts per second per square metre, and ω is the field of view in sr, then

$$sA_xT = 3(B_iT + B_cA_pT + B_dA_x\omega T)^{\frac{1}{2}}. \tag{4.1}$$

If now we provide some system of efficiency ε to label particle counts, so they may be rejected, where ε is the fraction of particle counts that escape labelling, we have

$$s = 3(B_i + B_c\varepsilon A_p + B_dA_x\omega)^{\frac{1}{2}}/A_xT^{\frac{1}{2}}. \tag{4.2}$$

We notice that there is a critical field of view which, by inserting the typical values of B, gives

$$\omega_c \sim 0·1\varepsilon A_p/A_x \text{ sr}. \tag{4.3}$$

For values of field of view ω larger than ω_c the diffuse background dominates the noise, and below this the particle flux dominates it. A_p and A_x are normally not very different. If we equate them,

$$\omega_c \sim 10^{-2}\varepsilon \text{ sr}. \tag{4.4}$$

For $\varepsilon = 1$ (no background rejection) $\omega \sim 5°$ square, and for $\varepsilon = 0·01$ (about as good as currently attainable) $\omega \sim 0·5°$ square. These cover the range of fields normally obtained by honeycomb or slat (egg-crate) collimators.

Eqn. (4.2) shows that, in the diffuse background-dominated mode, the minimum detectable signal varies inversely as the root of the sensitive area of the system, while in the particle-dominated mode the minimum signal varies inversely as the area. It is clear therefore that increasing the sensitive area is more rewarding for small fields of view than for large. In every mode the minimum detectable signal varies inversely as the root of the observation time.

We will now compare the behaviour of a large-area counter with that of a detector at the focus of a reflecting telescope (see later), in which the ratio of X-ray to particle collecting areas can take a very large value. This fact is very useful when the field of view of the telescope is small enough to take advantage of the reduction in background, but it makes a telescope of little use for providing a survey with a resolution of, say, 0·5°. At this field of view the minimum detectable signal s_c for a counter alone, with $\varepsilon = 0·01$ is, from eqn (4.2),

$$s_c \sim 18(A_x T)^{-\frac{1}{2}}, \tag{4.5}$$

differing little from

$$s_t \sim 12(A_x^* T)^{-\frac{1}{2}}, \tag{4.6}$$

which applies to a small counter at the focus of a telescope having an area a hundred times that of the counter. Here the effective area of the telescope for signal collection has been written A_x^*, and the area A_x of the unaided counter sensitive to X-rays has been equated to its area A_p sensitive to particles.

Moreover, to make a telescope with a given effective area requires far more weight and is much more difficult than to make a counter of twice the effective area, and so for large-scale survey purposes the latter is chosen. However, if we compare a telescope with a much smaller field of view, say, 1′ of arc square, with a counter having a 0·5° square field (which it would be difficult to improve on), we obtain

$$s_c = 18(A_x T)^{-\frac{1}{2}}, \tag{4.7}$$

compared with

$$s_t = 1·4(A_x T)^{-\frac{1}{2}} \tag{4.8}$$

where the telescope A_p has been taken as $0·1\, A_x^*$.

In this case, $A_x^* = (18/1·4)A_x$. That is, the effective telescope area for the same minimum detectable signal would be 0·08 that of the counter. We shall return later to consider the factors that determine the effective area of the telescope, and we note that the argument is affected if the detector at the focus of the telescope is not simply a whole-field photometer but is a position-sensitive system. The above calculations, however, serve to demonstrate one reason why the decade of initial sky-surveying was involved with mechanically-collimated counters rather than focusing systems.

A multi-wire proportional counter

As an illustration of the techniques used to obtain a small value of ε, that is, to get good discrimination against charged particle counts, we shall describe a counter prepared for flight on UK-5 (Ariel 5). Although this is a satellite system, it represents the kind of technique employed in contemporary rocket payloads; but it must be emphasized that early flights were carried out with less sophisticated arrangements.

In general, three methods may be used to discriminate against charged particle background:

1. The detector may be surrounded, except for the window, by a system sensitive to the passage of charged particles. For small detectors at the foci of telescopes a crystal of sodium iodide viewed by a photomultiplier is suitable. A large, flat detector array may be backed by a similar window-less array, which may share the same chamber and gas filling. Since very energetic particles frequently penetrate the whole arrangement an anti-coincidence array on the back picks up most particles entering through the front.

2. The shape of the pulse produced at the anode is, in general, different for X-rays than for charged particles. A pulse rise-time discriminator circuit can filter out 90 per cent or so of the unwanted pulses.

3. The counter is built of a number of cells, so connected that, if the long track of a charged particle causes a count in more than one cell, the count is rejected by the anti-coincidence circuit. When all three systems are used together it is possible to obtain values of ε of 10^{-2} or less.

The counter depicted in Fig. 4.5 makes use of all three systems, though pulse-shape discrimination is more a matter of circuit design than of counter geometry. The upper part of the diagram is a section of the counter, showing that it consists of a gas-containing vessel carrying a beryllium foil window surmounted by a honeycomb collimator, which both supports the window against the pressure differential arising from the vacuum of space and provides a field of view of about 5° diameter. The detector array consists of 32 long cells of square cross-section, formed by stretching fine, earthed wires between frames. Small insulating inserts in the frames, seen in the centre of each cell in Fig. 4.5(a), provide a means for stretching a fine anode wire down the centre of each cell. The cells are connected to four pairs of amplifiers in the way shown in Fig. 4.5(b). If a signal is received in more than one amplifier it is rejected. In addition, the array is surrounded on three sides by a further set of guard cells wired to a single amplifier, such that coincident signals between it and any other amplifier result in rejection.

Binary mask systems

An evident objection to mechanical collimators of the honeycomb type for exploratory surveys comes from the fact that no information is collected except

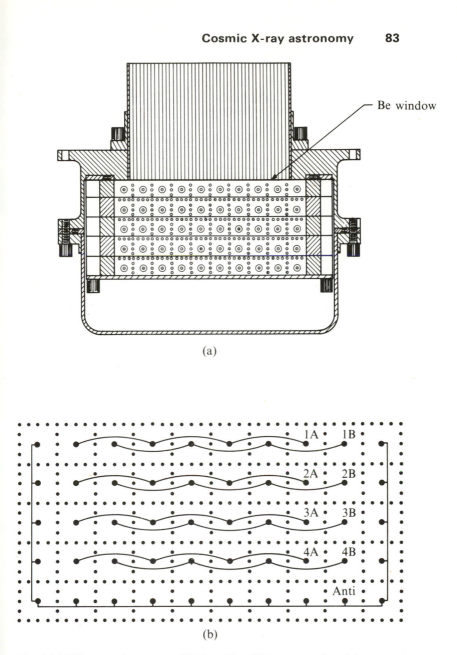

(a)

(b)

Fig. 4.5. (a) Diagram of counter on UK-5 satellite. (b) Interconnection of the counter cells showing how a long-track ionizing particle may be discriminated against (Mullard Space Science Laboratory).

when an object is in the field of view. Fig. 4.6 represents one method of over-coming this objection. A pair of grids is mounted in front of the detector with their axes parallel to the spin axis of the rocket. The field of view is now divided into numerous planes, with the maximum transmission normal to the detector. By making the grids very fine, a resolution of a few minutes of arc can be achieved, although photons are collected over a much wider field. A best fit is made of the measured signal to the response of the system as a function of roll angle.

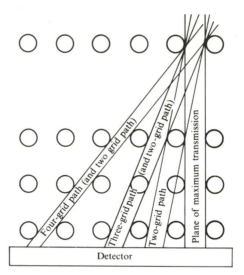

FIG. 4.6. Grid collimator.

The narrower the grid pitch the greater is the possibility of uncertainty about which maximum in the signal corresponds to the normal to the detector. To remove this uncertainty one or more additional pairs of grids may be in-serted, as shown in the figure, to eliminate response in some of the planes. It should be emphasized that the figure grossly exaggerates the grid pitch and that, in practice, the engineering problems of making and maintaining constancy of grid pitch and parallelism of the grids are difficult and set a limit on the resolution which can be achieved.

The arrangement just described only gives good position data in one plane and, although it is possible to sort out the signals from more than one source in the field at one time, it is not very useful on extended sources. An improved arrangement, known as a *rotation modulation collimator*, employs a double grid system mounted normal to the axis of a spacecraft. Such a system is illustrated in Fig. 4.7. As the spacecraft rotates, the signal from a source is

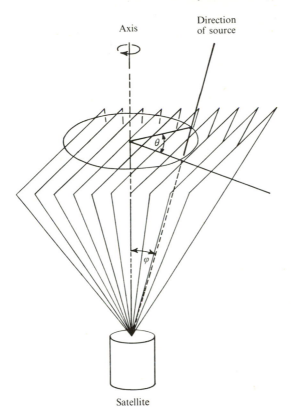

FIG. 4.7. Action of a rotation modulation collimator.

modulated by the transmission pattern of the grid, such that there is a minimum frequency of modulation while the source lies in a plane through the
rotation axis normal to the plane of maximum transmission, and a maximum
frequency while the source lies in that plane.

Fig. 4.8 illustrates the response of the grid array fitted to UK-5 as obtained
during a laboratory calibration with a white-light source 9° off-axis. Again it

FIG. 4.8. Signal generated by the rotation modulation collimator on Ariel I during
laboratory calibration with a single off-axis source (Mullard Space Science
Laboratory).

transpires that the signals from several sources in the field at once can be unscrambled, and in this case full position data (both θ and ϕ) is obtained. The theory of the coding and decoding is analogous to that of the Fourier transform.

We shall see later that reflection optics for X-rays only work efficiently in the soft part of the spectrum. It may well be that beyond 5 keV the immediate future of X-ray astronomy will depend on binary masks to combine a wide field of view and a large collecting area with a relatively high resolution. The two systems described so far are examples of a class which operate by encoding the position data in the form of a signal which is a function of time. A transformation is made between position and time, and the reverse transformation is made on analysis. (This may be compared to the way in which a television picture is dissected, transmitted as a time function, and reconstituted spatially.) It is possible, however, to devise binary mask systems which combine wide field, large area, and high resolution and yet encode the position data spatially. Of course, that is what a photographic camera does, but in that case there is a one-to-one relationship between light intensity observed in a given direction and signal recorded in the corresponding position. A hologram, on the other hand, records light intensity in a given direction by the intensity of a pattern all over the hologram, the pattern itself encodes the direction. This process too is comparable to the analysis of a periodic wave form into a Fourier series or a non-periodic amplitude distribution into a Fourier integral.

In one laboratory study of a typical binary system, a mask consisting of concentric opaque and transparent rings all of equal area (see Fig. 4.9(a)) provides the collecting aperture of the system. At some distance behind this is a sensitive photo-electric surface recording the shadowgram. After exposure, the distribution of shadows is reconstituted into an intensity map by scanning the plate with a very close electrode system of the shape illustrated in Fig. 4.9(b). When this array sits exactly above the rings of charge produced by a distant point source, the difference in potential induced in the two leads is a maximum. This is only one of many different ways of decoding the hologram-like record. It can be shown that the resolution of a system of this kind is about the thickness of the outer ring.

Fig. 4.10 shows the resolution in the laboratory of three sources separated by substantially less than the aperture of the rather crude five-ring 3 cm diameter array illustrated.

X-ray reflecting telescopes

It has been known for decades that X-rays can be reflected at glancing angles from smooth surfaces. The phenomenon is one of total external reflection, since the refractive index at X-ray wavelengths is just less than unity. Diffraction gratings used at very short wavelength are operated at grazing incidence—for this reason this technique has provided the basis for accurate measurement of X-ray wavelengths and so of finding crystal-lattice constants, Avogadro's

(a)

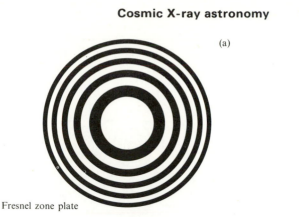

Fresnel zone plate

(b)

Electrode ring system

FIG. 4.9. A zone plate for encoding X-ray source positions as shadowgrams and an electrode system for searching for the shadowgrams recorded as a photo-electrically generated charge distribution. The systems were 3 cm in diameter.

number, and so on. The critical angle varies from substance to substance and is roughly proportional to the wavelength of the radiation. Important degradation in efficiency occurs near an X-ray absorption frequency (K, L, etc. edge) for the reflector, so that the material used has to be chosen with care and one material is not optimum over a very wide waveband.

A typical critical glancing angle would be about 1° at 2 keV (0·6 nm), though the reflection efficiency changes slowly as the angle is traversed, in contra-distinction to the behaviour of critical angles for glass or water in the visible. This indicates that transmission of X-rays in the reflecting substance is a lossy process (the refractive index is a complex number), which is well known.

The classical form of a simple reflecting optical telescope is a single parabola with a field stop and detector at its focus. An arrangement of this kind in which

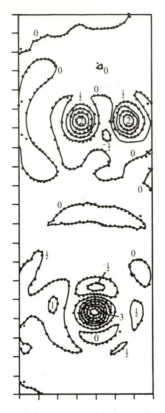

FIG. 4.10. Contour map made by scanning the charge shadowgrams of three sources made by the mask in Fig. 4.9 by the electrode system in the same figure. The scales are in millimetres showing that the resolution is a small fraction of the 3 cm aperture (Boyd, Booth, and Rawlings 1972).

the glancing angle ($\frac{1}{2}\pi$—angle of incidence) is kept small is illustrated in Fig. 4.11. The collecting area of this is a maximum when, the maximum glancing angle having been selected, the length of the reflecting element occupies half the total length L. It has, however, one serious limitation. It only focuses objects on the axis. A ray entering at a small angle α to the axis gives rise to an aberration pattern in the form of an annulus about the principal focus. A set of telescopes of this kind with tiny proportional counters at the focus constituted the first telescopes put in orbit for cosmic X-ray astronomy. They were built by the Mullard Space Science Laboratory of University College London, installed in the NASA Orbiting Astronomical Observatory launched in August 1972, and named *Copernicus* in honour of the 500th annervisary of that

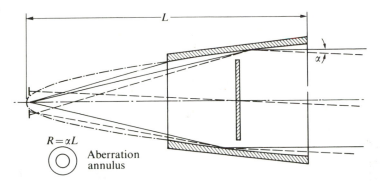

FIG. 4.11. Simple parabolic collector for X-rays showing annular form of image for off-axis rays. For maximum collecting area the mirror length is $L/2$.

astronomer's birth. There were three separate optical systems to cover the range from 0·2 nm to 6·0 nm.

A set of similar telescopes optimized for the soft end of the spectrum is installed on the satellite UK-6 scheduled for 1976; a Dutch satellite also carries a telescope of this kind, as will the European Space Agency satellite EXOSAT (Eccentric orbit Occultation SATellite). This satellite will be placed in an eccentric orbit and will use lunar occultation to improve positional information on known sources—a technique already used to good effect with sounding rockets and with the satellite *Copernicus*. Satellites for the more distant future, however, will use more sophisticated optics, capable of imaging a small area of sky and of transmitting the image without wasting photons in a process of sky-scanning or image-dissection. A satellite of this kind is the NASA High Energy Astronomical Observatory (HEAO-B). It will have a turret of interchangeable instruments at the focus, including position-sensitive detectors capable of recording pictures of the field, a solid-state detector using a cryogenically-cooled crystal of doped silicon, which gives better energy resolution than a proportional counter, and a Bragg crystal spectrometer.

Fig. 4.12 outlines the principles of the mirror system. By using two reflections, first from a parabola and then from a confocal, concentric hyperbola, the rays are brought to a focus at the other focus of the hyperbola. Although there are aberrations off axis such a system produces a true image.

By nesting several mirrors inside one another, good use may be made of the available dimensions, especially length, and the inner mirrors having a smaller glancing angle give higher efficiency at short wavelengths than the outer mirror. The effective collecting area is that area which with 100 per cent reflection would pass the same number of photons of a given wavelength through the focal area as does the actual mirror. This area decreases with

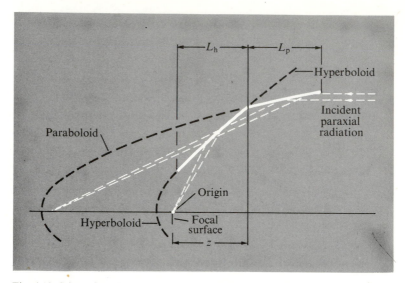

Fig. 4.12. Stigmatic pair of mirrors. L_p = length of parabolic section; L_h = length of hyperbolic section; Z = effective focal length. The external focus of the hyperbola is coincident with the parabola focus giving the final focus of the optics at the internal focus of the hyperbola. Angles of incidence in the diagram are grossly exaggerated.

decreasing wavelength, varying in the proposed HEAO telescope (optical diameter of largest mirror 0·56 m) from 0·05 m^2 at 0·3 nm to 0·42 m^2 at 4·4 nm.

Position-sensitive X-ray detectors

Although telescopes like this have not yet been orbited for cosmic X-ray astronomy they have been used with great success for solar work, for example, on Skylab (see Fig. 3.15). The picture shown in Fig. 3.15 was recorded photographically, the film being recovered by, and returned with, the astronauts. Such a facility is normally too costly, though during the 1980s this situation may well change. In any case, the fog background on a photographic film is a serious limitation on sensitivity in the X-ray part of the spectrum. A great deal of effort therefore has been put into producing position-sensitive detectors of X-rays with a resolution capability to match that of the telescope technology. Some of these have the additional advantage over photography of obtaining energy information from pulse size.

An instrument of this kind developed at the Mullard Space Science Laboratory for inclusion in several proposed payloads is represented in Fig. 4.13. It consists of a gas-filled proportional counter with planar geometry and with well-defined photon absorption and electron-multiplication regions

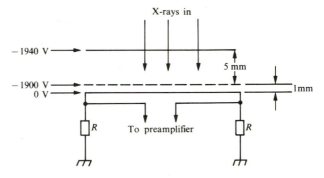

F<small>IG</small>. 4.13. Electrode system of a position sensitive X-ray proportional counter (Stümpel, Sanford, and Goddard 1973).

separated by a fine grid. Ionization produced by the photon in front of the grid results in electrons drifting towards it and, after passing through into the high-field region, being multiplied by a gas-ionization avalanche. The position of arrival of the charge pulse can be obtained by analysing the four current trains which arrive at four pick-off points around the periphery of the high-resistivity anode plate. The equations governing the flow of charge in the resistance–capacitance continuum of the plate are those representing the flow of heat dumped suddenly at a point in a disc having a given thermal conductivity and capacity.

This counter is intended for use at wavelengths out to about 6·0 μm, which necessitates use of a very thin window (~ 1 μm) of a substance such as polypropylene. Because thin windows of this kind are permeable to the filling gas, elaborate gas-storage and control-valve arrangements are necessary to maintain the density of the gas at a constant value.

Higher resolution, at the cost of loss of energy discrimination and some loss of photon sensitivity, may be obtained by replacing the gas photon-absorption and electron-multiplication arrangement by a mosaic of electron multiplier channels of the kind illustrated in Fig. 2.9. Channel multiplier plates of this kind can attain resolutions of around 10 μm. They are used in conjunction with a suitable photo-electric cathode on which the image is focused and the final picture can be read out in several ways, including that just described for the gas-filled counter.

The identification of X-ray sources

We saw earlier that, before the advent of satellites, both galactic and extra-galactic sources of unexpectedly high power were found; that the galactic sources included systems associated with optical objects (for example, Sco X-1) and supernova and pulsar remnants (for example, the Crab); that variability occurred on a wide scale (for example, Sco X-1, Cen X-2); and that there is an unresolved background. At the end of 1970 a great observational step forward was made with the launch of *Uhuru* (the word means freedom in the language of Kenya). The satellite was launched from a rig off the coast of Kenya at the latitude of the equator. This small, spinning satellite carried a large counter array with simple cellular mechanical collimators of the kind that had been widely used in rockets. The foil windows, which made it un-necessary to provide a top-up gas supply, limited the response to the photon energy range above 2 keV. The satellite carried out a crucial programme. It lead to the discovery of about 170 new sources (see Fig. 4.4), established approximate positions and obtained some broad spectral information on 200 sources, and provided data on variability and periodicity that only a satellite could obtain. Many of the astrophysical insights, to date, into the mechanisms of X-ray sources have been removed from the area of pure speculation by this satellite. Nevertheless, there remains a great deal of uncertainty. The greatest advance in understanding an X-ray source usually comes when it can be positively identified with a known optical or radio source. The positional accuracy obtainable by *Uhuru* is not normally adequate for these identifications, though correlation of periodic or secular behaviour can provide a strong pointer to identity. *Copernicus* has improved on posi-tional data to around 20″ of arc for bright sources, but it will probably be near the end of the 1970s before we can approach lunar occultation as a means of accurately establishing source positions (to 1″ of arc).

The astrophysics of X-ray sources

We shall describe briefly the current basic ideas underlying the energy source mechanisms for the three main classes of galactic X-ray source: supernova remnants, binaries, and pulsars.

It is interesting and most important for astrophysics and cosmology that all of these systems concern the end stages in the life of a star when its nuclear fuel is exhausted. A complex of mechanisms triggered by this situation can result in the sudden collapse of a star under the pull of its own gravity, no longer adequately counteracted by radiation pressure. In a supernova explosion of this kind the star shines extremely brightly for a short time, and a substantial part of its mass is shot out into space. The classical example of this is the Crab Nebula, formed in 1054 and recorded by the Chinese as visible in the daytime. The resultant nebulosity at a distance of about 4000 light-years subtends an angle a little over 1′ of arc. In 1964 this system became the first X-ray source (Tau X-1) to be identified with a visible object, as in 1949 it was the first ratio source to be so identified (Taurus A). Later, as we have seen, it was found to be a pulsar also, sending out pulses throughout the electro-magnetic spectrum at a frequency of 30 per second. This pulsing is thought to arise from the very rapid rotation resulting from the residual angular momentum of the stellar remnant, now condensed under its own huge gravitational pressure to the density of nuclear matter—a *neutron star*. (The high pressure forces the reaction equilibrium between protons, electrons, and neutrons primarily in the direction of neutron formation.)

The radius of a neutron star is of order 10 km. It appears likely that the energy for both the pulses and steady parts of the X-ray output of the Crab comes from rotation of the neutron star, the period of which can be measured very accurately by radio observations. It is found to slow down, with occasional *glitches*, as they are called, when it speeds up discontinuously, probably due to a change of mass distribution in a starquake. The pulsing is thought to be a lighthouse effect caused by a beaming of radiation resulting from an asymmetry in the star (presumably its magnetic field).

If we calculate the magnetic field likely to exist as the highly conducting medium of a star collapses in volume, by a factor of order 10^{15}, it appears that fields around 10^8 T may be expected. There is, around the star's axis, an imaginary cylinder known as the *velocity-of-light cylinder*, at which a vector fixed in the star would be moving at the velocity of light. For a rotational frequency f it is easy to see that this radius is given by

$$R = c/2\pi f, \tag{4.9}$$

which, for the Crab, is 1600 km. Somewhere inside this radius the electric fields due to the spinning magnetic field reach such a value as to result in acceleration of ambient electrons to energies high enough to generate radiation over a wide spectral range. Without establishing the details of the mechanism we can see that some kind of beaming of X-rays may be expected. We note in passing that the spectrum need not be that of a thermal source.

The form of the unpulsed spectrum suggests that it, like the radio emission from the bulk of the nebula, is due to the centripetal acceleration of electrons as they rotate in helices about the residual nebular magnetic fields (an idea confirmed by the discovery of polarization of the rays). This aspect of the production of radiation by centripetally-accelerated electrons (cf electrons in a radio aerial) is known as synchrotron radiation when the energies of the electrons are relativistic, since it occurs in electrons orbiting in a synchrotron machine. There remains the problem of the energy source for the radiating electrons. If the electrons obtained their energy only from the original explosion they would have cooled down long ago. The characteristic time for electrons radiating 1 keV X-ray photons is only 30 years in the fields of the nebular plasma ($\sim 10^{-8}$ T). It seems probable therefore that the rapidly rotating magnetic field of the pulsar not only produces the lighthouse beam but also, by its stirring action, heats up the whole plasma of the nebula.

The power radiated in X-rays is about 10^{31} W which, since the explosion, represents an energy of about 3×10^{41} J. The total kinetic energy of rotation of a neutron star 10 km in radius rotating at 30 revolutions per second is

$I\omega^2 \sim 10^{45}$ J, and so there would appear to be sufficient energy of rotation to account for the emission in this way.

Older supernova remnants which emit X-rays are not found to be pulsars. If they ever were, they have presumably dissipated much of their energy of rotation. An example of such supernovae is Cassiopeia-A, the strongest radio source in the sky. Here the *Copernicus* satellite has found the X-rays to correlate fairly closely with the radio flux, which appears to come from a shell-like structure, possibly a shock wave in the interstellar medium, caused by the original explosion. However, Puppis-A, another old supernova remnant, shows no close correlation between radio and X-rays.

To the energy sources for X-rays discussed so far—nuclear energy in the Sun, rotational energy in the Crab, energy of a supernova explosion (that is, probably a combination of gravitational, nuclear, and rotational energy)—we must now add perhaps the most interesting of all, the huge gravitational energies released in *binary star* systems which emit X-rays. These systems were discovered when stars whose optical spectra showed periodic Doppler shifts associated with orbital motion around a shared centre of gravity were identified with X-ray sources and when some of the X-ray emitters were found to undergo periodic occultation. A substantial number of the X-ray sources in our galaxy can be accounted for by the accretion of matter from a normal hot star by the great gravitational potential of a neutron star, or similar condensed object, in an orbit around their common centre of gravity. This is true, at any rate as far as energy release is concerned, though the proposed detailed mechanisms vary considerably and depend greatly on the angular velocity of the condensed object and the presence or otherwise of a magnetic field.

We have already seen that the Sun emits a strong solar wind. In hot, young stars this wind can exceed 10^{-7} solar masses a year. A neutron star typically has a mass about equal to the Sun 2×10^{30} kg. If we suppose that such an object is in a very close mutual orbit around a hot star it might easily capture 10^{-10} solar masses a year, 6.3×10^{12} kg s^{-1}. The gravitational potential at the surface of a neutron star of mass (M) 2×10^{30} kg and radius (r) 10 km would be given according to Newtonian mechanics (which amounts to neglecting the mass equivalent of the energy in a gravitational field—sufficient for an order-of-magnitude calculation) by

$$P = MG/r, \tag{4.10}$$

where

$G = 6.7 \times 10^{-11}$ N m^2 kg^{-2} (the gravitational constant),

so that

$$P = 1.34 \times 10^{16} \text{ J kg}^{-1}.$$

The energy release from the capture of 10^{-10} solar masses per year would be

$$W = 1.34 \times 10^{16} \times 6.3 \times 10^{12} \sim 10^{29} \text{ W},$$

which corresponds well with the kinds of energy involved. It is, of course, necessary to show that the energy is released mostly in the X-ray band. This involves calculating the temperature of the plasma produced ($\sim 10^8$ K), which involves study of a realistic model.

Additional evidence for this kind of concept of an X-ray source comes from sources which show not only an occultation period of days, due to rotation about the companion star, but also X-ray pulsar behaviour presumably due to rotation of the condensed object. One source of special significance must be mentioned—Cygnus X-1 (Cyg X-1). This shows the characteristics just described for a binary X-ray source, but when the mechanics of the observed motion are worked out it is found that the dense companion probably has a mass several times that of the Sun. Now it has been shown on the basis of our current knowledge of nuclear forces that an object of this mass, which has no *internal* energy source to inflate it, is unable to prevent its gravitational collapse beyond that of nuclear density. Indeed, as far as our present knowledge goes, such an object collapses indefinitely. In any case there will be a radius at which the escape velocity for a photon to leave the system exceeds the velocity of light. We can calculate the radius at which this will occur for a star of given mass. If m is the mass associated with a photon of given energy E, according to Einstein $E = mc^2$. But eqn (4.10) gives the Newtonian potential at a radius r, so the negative potential energy is

$$\frac{MG}{r}\frac{E}{c^2}.$$

For photon escape this must be less than E, that is,

$$M/r < c^2/G = 1\cdot34 \times 10^{31}.$$

Taking $M = 3$ solar masses $= 6 \times 10^{30}$ kg, the critical $r = 0\cdot45$ km. The correct relativistic term introduces a factor 2, giving $r = 2\,GM/c^2$. It is interesting that this result is easily obtained by calculating the radius at which the escape velocity of a particle is c, using Newtonian mechanics.

It now seems extremely probable that the massive, dense component of Cyg X-1 has shrunk beyond this limit and has become a *black hole*. While a black hole of this kind is indeed black, the region outside the critical radius is very far from black, being a plasma at millions of degrees. It is also important to realize that, although we can see no limit to the amount of matter a black hole can swallow, its gravitational potential at a distance is no different from that of an ordinary star of the same mass. Moreover, the rate of accretion is limited by radiation pressure from the hot plasma and by the need to disperse the angular momentum of in-falling material. If accretion is at the typical rate of 10^{-10} solar masses per year, it would take a time comparable to the age of the universe to double or treble the mass of the hole.

X-ray astronomy and cosmology

X-rays have been detected from individual galaxies, from clusters of galaxies, and from the quasar 3C273, which is thought to be at the huge distance of 2×10^9 light-years. If the distance is correct, 3C273 is the most energetic X-ray source known, radiating 10^{39} W. Our galaxy and its twin in Andromeda radiate 10^{32}–10^{33} W, but clusters of galaxies radiating an order of magnitude more power in X-rays per galaxy are known, and there is some evidence that much of this power may come from only a few of the members of the cluster.

These huge powers, which often greatly exceed the radio output (for example, by a factor of 100 for the radio galaxy M87) suggest that X-rays will be a vital tool in the exploration of the cosmos in the future. Quite apart from the understanding of the mechanisms and evolution of galaxies which may be expected, it is possible to see several ways in which X-ray observations will have a bearing on cosmology—our picture of the Universe and its history as a whole.

We have already seen that X-rays can provide a means for discovering condensed objects such as neutron stars and black holes, which might otherwise be undetected. By improving our estimate of the amount of matter in the Universe this can influence the expected distant future. As our knowledge stands, the Universe will go on expanding indefinitely, but if we find that much matter has been overlooked we may arrive at a gravitational potential ultimately sufficient to slow down and reverse the expansion. Even stronger evidence for (or against) additional matter may come from observations of X-rays produced in intergalactic material. Such data may point to an otherwise unobservable intergalactic gas at a temperature around 10^8 K. If this exists it would help us to understand also how the galaxies in the clusters themselves remain clustered.

Another matter whose ultimate resolution is likely to have important cosmological implications is the origin of the isotropic background. If it comes from an intergalactic medium, it would again affect our estimate of the mass of the Universe. It cannot be accounted for by the integrated radiation from distant normal galaxies like ours, which may imply that at the epoch of emission, so many light-years away, the galaxies were much hotter. In any case it may give us a glimpse of the Universe when it was younger and perhaps had a very different population of galaxies than at present.

It is interesting to find that a branch of astronomy that was impossible before the advent of spacecraft, and until 1962 met with some scepticism, is already making a very important (and may yet make a crucial) contribution to our understanding of the only Universe we have. However, here, as in the topics dealt with in earlier chapters, it is important to realize that science is a unity and that, important and exciting as the science done with spacecraft is, it is not a separate discipline but just another window through which we may look out on our world.

References

BOWEN, P. J., BOYD, R. L. F., HENDERSON, C. L. and WILLMORE, A. P. (1964). *Proc. R. Soc.* A**281**, 514, 526.

BOYD, R. L. F., BOOTH, R. F. and RAWLINGS, R. C. (1972). *J. Phys.* E **5**, 808.

—— LAFLIN, S. (1968). *Proc. R. Soc.* A**307**, 449.

BURTON, W. M., RIDGELEY, A. and WILSON, R. (1967) *Mon. Not. R. astr. Soc.* **135**, 207.

CHAPMAN, S. (1931). *Proc. phys. Soc.* **43**, 26.

CULHANE, J. L., SANFORD, P. W., SHAW, M. L., PHILLIPS, K. J. H., WILLMORE, A. P., BOWEN, P. J., POUNDS, K. A. and SMITH, D. G. (1969). *Mon. Not. R. astr. Soc.* **145**, 435.

FÖPPL, H. *et al.* (1967). *Planet Space Sci.* **15**, 357.

GIACCONI, R. and GURSKY, H. (1965). *Space Sci. Rev.* **4**, 151.

STRAUSS, F. M. and PAPAGIANNIS, M. D. (1971). *Astrophys. J.* **164**, 369.

STÜMPEL, J. W., SANFORD, P. W. and GODDARD, H. F. (1973). *J. Phys.* E **6**, 397.

VAIANA, G. S., DAVIS, J. M., GIACCONI, R., KRIEGER, A. S., SILK, J. K., TIMOTHY, A. F. and SOMBECK, M. (1973). *Astrophys. J.* **185**, L47.

VAN ALLEN, J. A., LUDWIG, G. H., RAY, E. C. and MCLLWAIN, C. E. (1961). *Annls int. geophys. Year* **12**, 671.

WRENN, G. L. (1969). *Proc. IEEE* **57**, 1072.

—— WILLMORE, A. P. and BOYD, R. L. F. (1962). *Planet. Space Sci.* **9**, 765.

Further reading

THE following books provide further reading, generally at a more advanced level and contain extensive references and bibliography.

Physics of the earth's upper atmosphere. (1965) edited by C. O. Hines, I. Paghis, T. R. Hartz and J. A. Fejer. Prentice-Hall, New York.

An introduction to the ionosphere and magnetosphere. (1972) by J. A. Ratcliffe. Cambridge University Press (see also *Sun, earth, and radio* (1970), by the same author, intended for the general reader, published by World University Library).

Introduction to solar terrestrial relations. (1965) edited by J. Ortner and H. Maseland. D. Reidel Publishing Company, Dortrecht.

Space Physics. (1970) by R. Stephen White. Gordon and Breach, London and New York.

None of the above books deals with cosmic X-ray astronomy.

Space physics and space astronomy. (1972) by Michael D. Papagiannis. Gordon and Breach, London and New York.
X- and gamma-ray astronomy. (1973) edited by H. Bradt and R. Giacconi. D. Reidel Publishing Company, Dortrecht.
Scientific satellites. (1967) by William R. Corliss. NASA, Washington D.C.

Index

adiabatic invariant, 33
ambipolar field, 25, 36
Andromeda, 96
Appleton, 1, 2
Ariel 1, 18 *et seq.*, 48, 49
Aurora, xv, 26, 38 *et seq.*, 47, 49, 67

Balfour Stewart, 1, 26
binary mask, 79, 82 *et seq.*
binary star, 92, 94
black hole, vii, xiv, 95, 96
Briet, 1
Bremsstrahlung, 63

C-layer (region), 2, 8, 9
Cassiopeia-A, 94
Cen X-2, 77, 92
Chapman, 4 *et seq.*, 26
chromosphere, 7, 55, 57, 59, 63, 65, 67, 72
conjugate points, 39
convection zone, 55
Copernicus satellite, 88, 92, 94
corona, xiv, 2, 7, 37, 54 *et seq.*, 61, 65, 72
coronagraph, 56, 62, 73
cosmic rays, xv, 9, 28 *et seq.*, 80
Coulomb field (force), xiii, xvi, 20, 36
counter, 64, 68, 75, 76, 79 *et seq.*, 90, 91
Crab nebula, 75, 77, 91 *et seq.*
critical frequency, 3
current dumping, 18, 23
Cyg X-1, 95

D-layer (region), 2, 8, 9
Debye (length), xiv, 15, 16, 25
diffuse X-ray background, 79 *et seq.*, 96
dynamo (action), xv, 1

E-layer (region), 2, 3, 7 *et seq.*
electric fields (in space plasmas), xiii, xiv, 25, 38, 41, 46 *et seq.*
ESRO-4, 25
exosphere, 36
Explorer I, 28, 29
Explorer III, 30
Explorer XXXI, 24, 25

F-layer (region), 2, 3, 8, 11, 12, 21
flare, 7, 56, 63, 65 *et seq.*, 75
floating potential, 16, 17
fluxgate magnetometer, 45, 46

galaxy, xiv, 78, 91, 96
geopotential metre, 4
GEOS, 41
Giacconi, 72, 75, 76
granulations, 54

HEAO-B, 89, 90
Heaviside, 1, 2
hologram, 86

intergalactic medium, xiv, 96
International Geophysical Year, 28
interplanetary medium, xiv, 96
interstellar medium, xiv, 74
ionic composition, 4, 25 *et seq.*
ionogram, 2 *et seq.*
ion temperature, xiv, xv, 13, 16, 22 *et seq.*

Kennelly, 1, 2

Langmuir, xiii, 16 *et seq.*, 23, 41

magnetic
  mirror, 31, 33, 36, 47, 50
  storm, 35, 37, 39 *et seq.*
  variations, 1, 10
magnetohydrodynamics, 36, 37, 47
magnetometer, 44 *et seq.*
magnetosheath, 37
magnetosphere, vii, xvi, 16, Chapter 2, 53, 65
Marconi, 1
Mariner IV, 12
Mariner V, 12, 13
Mars, 12
mass spectrometry, 9, 10, 21 *et seq.*
meteor, xv, 10
milky way, 78

negative ions, 8, 9
neutral atmosphere, 4 *et seq.*